Groin Pain Syndrome

Raul Zini • Piero Volpi
Gian Nicola Bisciotti
Editors

Groin Pain Syndrome

A Multidisciplinary Guide to Diagnosis and Treatment

Springer

Editors
Raul Zini
Department of Orthopedic Surgery
Maria Cecilia Hospital
GVM Care and Research
Cotignola
Ravenna
Italy

Piero Volpi
Knee Orthopaedic and Sport Traumatology
Humanitas Research Hospital
Rozzano
Milan
Italy

Gian Nicola Bisciotti
Kinemove Rehabilitation Center
Pontremoli
Massa Carrara
Italy

ISBN 978-3-319-41623-6 ISBN 978-3-319-41624-3 (eBook)
DOI 10.1007/978-3-319-41624-3

Library of Congress Control Number: 2017939588

Printed on acid-free paper

This Springer imprint is published by Springer Nature
The registered company is Springer International Publishing AG
The registered company address is: Gewerbestrasse 11, 6330 Cham, Switzerland

Foreword

Groin pain is a common complaint; notorious for the discomfort and disability it causes, it represents a diagnostic challenge for the orthopaedic surgeon.

In clinical practice, we are frequently faced with this problem, especially when treating athletes participating in sports that require repetitive lower-limb movements, kicking, and sudden changes of speed and direction during the sporting activity.

To better understand this challenging multifactorial condition, a comprehensive approach is mandatory. Patients often refer to vague and diffuse symptoms that can be attributed to a variety of pathological entities; hence, the diagnosis and management of the underlying pathological processes remains a challenge. Furthermore, the pain is frequently not attributable to a single etiological cause, but is derived from the possible association of multiple pathologies, thus requiring a differential diagnosis. The surgeon should carefully consider the differential diagnosis, as it is also important to take into account that groin pain can have musculoskeletal as well as visceral causes.

This book, representing the latest updates of current knowledge on this subject, is an indispensable resource that will assist the reader to recognize and treat groin pain. The reader will also become acquainted with many aspects of basic science that have improved the differential diagnosis and treatment of this syndrome.

The contributing authors are experts in the field and share with the reader their experiences and surgical insights. This book is a state-of-the-art guide for those starting in the field of groin pain syndrome, as well as for experienced specialists.

Rome, Italy Vincenzo Denaro

Preface

Groin pain syndrome is an important current clinical topic in sports medicine, but the syndrome is still not well known and its characteristics are debatable, while its patterns of diagnosis and treatment are often imprecise and fragmentary.

The classification and identification of a diagnostic taxonomy for the syndrome have not yet found a common interpretation. Therefore, gaining actual agreement among the different specialists that face the clinical issues of this entity is often difficult.

The terminology used in the literature includes, often inappropriately, a very large number of terms and definitions. Therefore, there is a need to organize and better classify the numerous clinical symptoms that can constitute groin pain; only in this way will it will be possible to reach a correct diagnosis and then provide the correct treatment.

This syndrome covers a range of symptoms, localized in the groin area, that often have a multifactorial etiopathogenesis, and as a consequence, treatment will involve practitioners from different medical specialties. The syndrome provides a clear example of how interaction and cooperation among different professional approaches, sharing patterns of diagnosis and procedures, may be fundamental in improving treatment.

This book aims to summarize the state of the art in the treatment and diagnosis of groin pain syndrome in an innovative way, considering the set of problems from different professional points of view and searching for a shared synthesis.

In the first chapter, a new classification of groin pain syndrome is suggested for a more precise differential diagnosis; this is based on the results of the recent Consensus Conference, held in Milan on February 5, 2016, with the participation of major Italian experts in the field.

The groin area is extremely complex, both from the anatomical and biomechanical points of view, and these topics will be analyzed, together with radiological diagnoses, to complete the introductory part of the book. The next part of the book describes orthopaedic pathologies that can cause groin pain; the different pathologies are examined both in terms of the diagnosis and in terms of the relevant treatment; we focus on labrum lesions; chondral damage; and pathologies of the central, peripheral, and peritrochanteric compartments.

The main pathology that causes groin pain is certainly femoroacetabular impingement, and above all, cam impingement, where a femoral bump leads to an articular conflict between femur and acetabulum, with associated biomechanical alteration; this produces compensatory stresses in other anatomical areas, such as the pubic symphysis, the sacroiliac joint, and the abdominal lining.

Also, groin pain can often be caused by visceral pathologies that are very important to understand and evaluate.

The so-called sports hernia is quite frequent in sportsmen, and is often not well diagnosed or well treated. Therefore, there is a need for correct diagnostic and therapeutic approaches; these are explained in Part II of this book.

Many tendon and muscle pathologies are also a frequent cause of groin pain, and in Part III of this book, these are discussed in depth, with particular attention paid to the adductor tendons, ileopsoas tendons, and abdominal muscles that can frequently be involved in atypical groin pain.

The pubic bone, both at the symphyseal joint and in the bone itself, can be the target area of compensatory stress, and therefore all possible pubic pathologies are examined.

Furthermore, we must never underestimate the neurological causes of groin pain, such as those that are well identified in the so-called nerve entrapment syndrome; these causes are discussed and investigated.

The end of the book contains chapters dedicated to rehabilitation; conservative rehabilitation is discussed above all, with particular attention given to strengthening and stretching exercises, physical therapy, and hydrotherapy.

With regard to postoperative rehabilitation, different chapters will discuss rehabilitation after arthroscopic hip surgery, abdominal surgery, and tendon surgery, with indications and in-depth analysis of the most up-to-date and efficient protocols.

I believe that, in the long run, this book will make a complete and thorough contribution to a multidisciplinary approach to the groin pain syndrome, and may therefore help specialists in different fields to better understand the various and often overlapping causes of groin pain.

The scientific co-operation of distinguished international and Italian experts, whom I kindly thank for their important contributions, has allowed us to reach our target; that is, to define the state of the art in our knowledge of the groin pain syndrome and then to suggest new diagnostic, therapeutic, and organizational models for the condition.

Cotignola, Ravenna, Italy Raul Zini

Contents

Part 1

Introduction

Classification and Differential Analysis of Groin Pain Syndrome

1

Gian Nicola Bisciotti and Piero Volpi

1.1 Introduction

It is never banal, before tackling a subject so controversial as the groin pain (GP), to remember that with this term—as on the other hand also as it regards all the other terms as athletic groin, groin disruption, osteitis pubis, etc. expressing the same kind of symptoms—we mean only the description of a symptom or better of a cohort of symptoms that the patient complains at the level of the pubic area. For this reason, we should be extremely careful not to identify the term GP (or any other term up to now considered as equivalent) and diagnosis. In fact, the GP has a multifactorial pathogenesis where often different clinical frameworks overlap, making sometimes the diagnosis a real diagnostic challenge. Objectively, it must be acknowledged that the anatomical complexity of the pubic region certainly does not facilitate the adoption of a clear nosological terminology. The multiple anatomical structures that may be involved are so numerous as to preclude, in fact, a comprehensive nomenclature [1], unless, as already pointed out, the term of GP is intended only as a description of a cohort of symptoms and not misunderstood with the diagnosis itself. Indeed, unfortunately, this simple and basic concept seems to be often overlooked in the specific literature. This lack of clarity uniqueness concerning terminology can be explained, but not of course justified, by the fact that since the symptoms reported by the patient can result from skeletal muscular, gastrointestinal, urogenital, neurological and gynaecological problems [2, 3], the risk for the clinician to use different terminologies is high. The fact is that the terminology is often confusing and sometimes dichotomous, a situation that creates a lot of difficulties of interpretation. Furthermore, there are objectively considerable difficulties in finding and interpreting the results reported by various studies. In this regard, a paradigmatic example is provided by Serner et al. [4] that in their systematic review emphasizes the need to standardize the terminology used in order to facilitate the comparison of results derived from the different studies present in literature. Not surprisingly, to reinforce this need, in their review, the authors included 72 studies, in which they found 33 different diagnostic terms. Recently, the "Agreement Meeting on Definitions and Terminology on Groin Pain in Athletes" held in Doha (Q) in November 2014 [5] was aimed to standardize the clinical terms used for GP.

G.N. Bisciotti (✉)
Qatar Sport Medicine and Orthopaedic Hospital, FIFA Center of Excellence, Doha, Qatar
e-mail: bisciotti@libero.it

P. Volpi
Istituto Clinico Humanitas, Rozzano, Milano, Italy

Responsabile Settore Medico F.C. Internazionale, Milano, Italy

R. Zini et al. (eds.), *Groin Pain Syndrome*, DOI 10.1007/978-3-319-41624-3_1

1.2 The Groin Pain Syndrome Italian Consensus Classification

The first Groin Pain Italian Consensus held in Milan February 5, 2016, was an invitation consensus conference attended by orthopaedics, sports physicians, general surgeons, radiologists, physiatrists, sport physiologists, physiotherapists and physical trainers that was aimed to approve three separate documents concerning the GP:

1. Diagnostic taxonomy document consensus
2. Clinical semeiotics document consensus
3. Imaging document consensus

Each document was first presented by a facilitator; the presentation was then followed by a plenary discussion directed by a chairman. After each discussion followed a vote. The first document has required 15 different discussions and the same number of votes, while the second and the third document required six discussions and votes. During the discussions, the document was eventually changed and was then voted only the final version. All votes are passed unanimously.

1.2.1 Summary of the First Document: Diagnostic Taxonomy Document Consensus

The first vote concerned the use of the term groin pain syndrome (GPS). The use of the term "syndrome" is justified by the frequent overlapping of different clinical frameworks and by the possible cause-effect interaction that characterize a well-defined GP clinical framework [6–8]. Obviously, the term GPS is an "umbrella term" that must necessarily be complemented by the clinical framework description. You may then, for example, have a GPS caused by adductor tendinopathy, or from inguinal hernia, or by a combination of these as of other pathologies. Therefore, it is our opinion that only adopting a comprehensive descriptive term, as GPS, and associating it with the clinical and taxonomic description of the disease, or diseases, responsible for the symptomatology reported by the patient, we can arrive to have a clear and rational identification of the problem. Then it was then proposed and approved the following definition of GPS:

> *Every clinical situation complained by the patient at the level of the inguinal-pubic area that affects the sporting activities and/or interferes negatively in activities of daily living (ADL) and requiring medical attention*

Furthermore, based on the synthesis of different studies [5, 9–17], we propose that the clinical frameworks that can be the cause of occurrence of GPS can be subdivided into 11 different categories as follows:

1. **Articular causes**
   (a) Acetabular labrum tear
   (b) Femoroacetabular impingement(I)
   (c) HALTAR lesion(II)
   (d) Hip osteoarthritis
   (e) Intra-articular loose bodies
   (f) Hip instability
   (g) Adhesive capsulitis
   (h) Legg-Calvé-Perthes disease and its outcomes
   (i) Dysplasia and its outcomes
   (j) Epiphysiolysis and its outcomes
   (k) Avascular necrosis of the femoral head
   (l) Sacroiliac joint disorders
   (m) Lumbar column disorders
   (n) Synovitis

   Notes:
   (I) Cam-Fai, Pincer-Fai, subspine impingement (or AIIS, anterior inferior iliac spine impingement).
   (II) Hip anterosuperior labral tear with avulsion of the rectus femoris.

2. **Visceral causes**
   (a) Inguinal hernia(I)
   (b) Other types of abdominal hernia
   (c) Intestinal diseases
   Note:
   (I) Concerning inguinal hernia, it is recommended to adopt the classification proposed by the European Hernia Society.
3. **Bone causes**
   (a) Fractures and their outcomes
   (b) Stress fractures(I)
   (c) Avulsion fractures(II)
   (d) Iliac crest contusion (hip pointers)(III)
   Notes:
   (I) Substantially concerning the pubic branch or the femoral neck.
   (II) Mainly the childhood avulsion fractures involving the anterior inferior iliac spine (AIIS), the anterior superior iliac spine (ASIS) and the apophyseal nucleus of the ischial tuberosity (ANIT).
   (III) The iliac crest contusion or hip pointers are the result of direct trauma at the level of the iliac crest which causes the formation of a periosteal haematoma. Such a haematoma can compress the lateral nerve femoro-cutaneous nerve and cause paresthesia symptoms.
4. **Muscle-tendon causes**
   (a) Rectus abdominis injuries
   (b) Rectus abdominis tendinopathy
   (c) Adductor muscles injuries
   (d) Adductor tendinopathy
   (e) Rectus abdominis—adductor longus common aponeurosis injuries
   (f) Iliopsoas injuries
   (g) Iliopsoas tendinopathy
   (h) Other indirect muscle injuries and their outcomes
   (i) Direct muscle injuries
   (j) Iliopsoas impingement(I)
   (k) Snapping internal hip
   (l) Snapping external hip
   (m) Bursitis(II)
   (n) Weakness of the inguinal canal wall(III)
   Notes:
   (I) Iliopsoas impingement with the medial portion of the acetabular rim.
   (II) Substantially concerning of the ileo-pectineal bursa and the greater trochanter sero-mucous bursa.
   (III) It's important to underline the four most important clinical signs of the inguinal canal wall weakness: tenderness to the exploration of the inguinal canal, tenderness on palpation at the level of the pubic tubercle, superficial inguinal ring dilatation and pain on palpation at the level of origin of the adductor muscles. In addition, an anamnestic index of extreme importance is a history of failure of conservative treatment.
5. **Pubic symphysis-related causes**
   (a) Osteitis pubis
   (b) Symphysis instability(I)
   (c) Symphysis degenerative arthropathy
   Note:
   (I) the radiological sign of symphysis instability is represented by an asymmetry of pubic branches greater than 3 mm visible in the Flamingo view X-ray.
6. **Neurological causes(I)**
   (a) Nerve entrapment syndrome(II)
   Notes:
   (I) The category "neurological causes" should be divided into two further subcategories. In the first category, they are the neurological damage due to overloading or overstretching (neurological causes category A). In the second category, they are the neurological damage due to an acute compression mechanism or tear of nerve structure (neurological causes category B).
   (II) Substantially concerning the femoro-cutaneous nerve, genitofemoral nerve

(genital branch), ilioinguinal nerve, iliohypogastric nerve, femoral nerve and obturator nerve.

7. **Developmental causes**
   (a) Apophysitis(I)
   (b) Growth plate at pubic level(II)
   Notes:
   (I) Substantially concerning the AIIS and the ASIS.
   (II) Below the age of 20 years is common to observe anteromedial foci of endochondral ossification centres. These findings become particularly evident in arthro-IMR [18].
8. **Genitourinary disease-related causes (inflammatory and not)**
   (a) Prostatitis
   (b) Epididymitis
   (c) Funiculitis
   (d) Orchitis
   (e) Varicocele
   (f) Hydrocele
   (g) Urethritis
   (h) Other infections of the urinary tract
   (i) Cystitis
   (j) Ovarian cysts
   (k) Endometriosis
   (l) Ectopic pregnancy
   (m) Round ligament entrapment
   (n) Testicular/ovarian torsion
   (o) Ureteral lithiasis
9. **Neoplastic causes**
   (a) Testicular carcinoma
   (b) Osteoid osteoma
   (c) Other carcinomas
10. **Infectious causes**
   (a) Osteomyelitis
   (b) Septic arthritis
11. **Systemic causes**
   (a) Inguinal lymphadenopathy
   (b) Rheumatic diseases

After a deep examination and discussion concerning the literature, we propose to subdivide the most common and probable diseases that can cause GPS in 11 different nosological categories including 63 possible different clinical frameworks (Table 1.1).

Table 1.1 The most likely causes of GPS (63) grouped into 11 different nosological categories

Categories		Number of pathology
Articular causes		14
Visceral causes		3
Bone causes		4
Muscle-tendon causes		14
Pubic symphysis-related causes		3
Neurological causes		1
Developmental causes		2
Genitourinary disease-related causes (inflammatory and not)		15
Neoplastic causes		3
Infectious causes		2
Systemic causes		2
Total	**11**	**63**

Into the last part of the first document, the consensus approved a further subdivision of the GPS in three main categories, based both on the aetiopathogenesis and the timing of onset/disappearance of the clinical framework:

1. The GPS of traumatic origin, in which the onset of pain was due to a precise traumatic event and this hypothesis is supported by the anamnestic investigation, by clinical examination and imaging.
2. The GPS due to functional overload, characterized by insidious and progressive onset, in which the patient has no memory of trauma or a situation to which is attributed with certainty the onset of pain symptoms.
3. The long-standing GPS (LSGPS) or chronic GPS, in which the cohort of symptoms complained by the patient continues for a long period and is recalcitrant to any conservative therapy. It's important to underline the fact that both the functional overload GPS and the traumatic origin GPS may hesitate in a LSGPS. Similarly, a traumatic GPS can occur in a previous framework of GPS by overuse and/or LSGPS. We can consider in this category a clinical framework that has continued for more than 12 weeks. Finally, it is interesting to underline that a situation of

LSGPS is typically most commonly encountered in an amateur athlete rather than in a professional athlete. This can be reasonably explained by the fact that an amateur athlete does not have the same access opportunities to a professional athlete to have a suitable therapeutic procedure, either conservative or surgical.

Therefore, a correct formulation of the diagnosis, corresponding to the concepts stated above, should respect the following formulation: "traumatic GPS caused by..." or "overuse GPS caused by..." or "LSGPS caused by …".

Finally, we underline the concept that, given the anatomical complexity of the pubic region, especially the GPS due to functional overload and the LSGPS can often be caused by the association of more diseases. In the case of a type of GPS caused by the association of more diseases, the diagnosis formulation will change in "traumatic or overuse GPS, or LSGPS caused by the association of …".

1.2.2 Summary of the Second Document: Clinical Semeiotics Document Consensus

Before describing the second document concerning the semiotics, we would like to recall briefly the GPS cluster of signs and symptoms.

It is estimated that a percentage between 5 and 18% of athletes ask medical attention caused by an activity-restricting GPS [10, 19–21]. Within the same sport played, males had greater GFS incidence than females with a RR equal to 2.45 [5].

In the patient affected by GPS, the symptoms are bilateral in 12% of the cases; it involves the adductor region in 40% of the cases and the perineal region in 6% of the cases: The symptom usually begins unilateral and becomes with the progress of time bilateral [20–26]. The pain onset occurs insidiously in 2/3 of patients and acutely in the remaining 1/3, a certain number of patients refers an acute event after a clinical framework of GPS or LSGPS was already present [20, 22, 26–29]. The clinical framework is characterized by subjective and objective symptoms. Subjective symptoms are mainly represented by pain and functional deficits [30, 31]. From an objective point of view, the patient may complain pain on palpation, during countered muscle contraction and during passive and active stretching. The clinical examination must therefore be based on a series of tests focused on muscle contractions (isometric, concentric and eccentric), on the active and passive stretching manoeuvres [32–36] and on the palpation of some specific anatomical areas [14, 37–40]. Thus, basing both on the examination of the literature and on expert opinion of the specialists present was approved a second document concerning the clinical examination. The clinical exams approved and recommended during the consensus were categorized in four categories as follows.

1.2.2.1 First Category: Specific Test for Abductor Muscles

1. Palpation of the pubic branch at common rectus abdominis/adductor longus common aponeurosis
2. Isometric squeeze test with proximal resistance (at knee level)
3. Isometric squeeze test with distal resistance (at ankles level)
4. Isometric squeeze test with distal resistance and apart legs
5. Isometric squeeze with flexed leg and proximal resistance
6. Isometric squeeze test in monopodalic execution with the use of a dynamometer(I)

Note:

(I) Optional test but in any case strongly recommended especially in the case of unilateral pain symptomatology.

1.2.2.2 Second Category: Specific Test for Abdominal Muscles

1. Palpation of the pubic branch at common rectus abdominis/adductor longus common aponeurosis

2. Rectus abdominis eccentric test
3. Sit-up pain test
4. Obliquus abdominis eccentric test

1.2.2.3 Third Category: Specific Test for the Hip Joint

1. Hip joint intra- and extra-rotation measurement
2. Flexion abduction external rotation (FABER) test
3. Dynamic internal rotatory impingement test (DIRIT)
4. Dynamic external rotatory impingement test (DEXRIT)
5. Posterior rim impingement test
6. Lateral rim impingement test

1.2.2.4 Fourth Category Clinical Evaluation of Inguinal Diseases

Palpation and clinical evaluation of the following anatomical structures:

1. Tuberculum pubis
2. Crista pubis
3. Linea pectinea
4. Superior ramus pubis
5. Anulus inguinalis superficialis
6. Pilastrum infero-lateralis
7. Pilastrum supero-medialis

Furthermore, as part of the second consensus document, it has approved the use, during the medical history process, of the HAGOS patient-reported outcome measures in its validated Italian form [41].

1.2.3 Summary of the Second Document: Imaging Document Consensus

The third document discussed and approved during the consensus involved the imaging exams. They were considered the protocols regarding the conventional radiology (X-ray), ultrasound examination (US) and magnetic resonance imaging (MRI). It was not made no division between first and second level exams, because it was considered that each exam has specific peculiarities. Therefore, basing both on the examination of the literature [11, 42–58] and on expert opinion of the specialists present was approved a second document concerning the imaging assessment which is composed by the following routine examinations:

1. *X-ray examination*

   The radiography routinely discussed and approved includes the following exams:
   (a) Anterior posterior view in upright position (AP1)
   (b) Anterior posterior view in upright position and alternately on one foot (Flamingo view) (AP2)
   (c) Dunn view (D)

   From the radiographic assessment, it is recommended to obtain the following information:
   (a) Presence of cross sign (AP1)
   (b) Enlargement and /or erosion and/or sclerosis of the symphysis (AP1)
   (c) Symphysis asymmetry greater than 3 mm (AP2)
   (d) Calculation of alfa angle (D)

2. *US examination*

   The US examination must provide the following assessments:
   (a) Assessment of the muscle-tendon unit of the abdomen and adductor muscles
   (b) Dynamic assessment for the inguinal canal structures
   (c) Assessment of internal organs
   (d) Assessment of the urinary tract and of the external genitalia

   Finally, during the execution of the US examination, the contemporary presence of the radiologist and the general surgeon is strongly suggested.

3. *MRI evaluation*

   Concerning the MRI evaluation, the use of a device of at least 1.5 T and a no-contrast protocol is recommended. The acquisition plans recommended are:
   (a) Coronal
   (b) Sagittal
   (c) Axial

(d) Axial oblique
(e) Coronal oblique
(f) Sagittal oblique

The acquisition sequences recommended are:
(a) T1
(b) T2 and T2 fat saturated (T2 FS)
(c) STIR
(d) Proton density fat saturation (PD FS)

Furthermore, into the third document, the consensus suggested the radiological findings of major interest:

The presence of bone marrow oedema (BMO) at pubis symphysis level. The presence of BMO must be identified into the sequences coronal STIR, coronal T1 and axial oblique T2 FS and PD FS. Furthermore, BMO must also be classified in I°, II° or III° in relationship of its extension measured into the PD FS or T2 FS axial oblique plan sequences.
Fatty infiltration within the BMO around the joint symphysis to verify into the coronal STIR, coronal T1 and axial oblique T2 and PD FS sequences.
Symphysis sclerosis to asses in coronal T1 and axial oblique T1 images.
High signal intensity para-symphysary line to verify in coronal STIR, axial oblique PD FS and sagittal STIR sequences.
Secondary inferior and/or superior cleft sign to assess in coronal STIR, axial oblique PD FS and sagittal STIR sequences.
Subchondral cysts/irregularities of the articular surface to verify in coronal STIR and axial oblique images.
Symphysis central disc protrusion to estimate in coronal T1 and axial oblique T1 sequences.
Adductor longus tendinopathy to assess into the axial oblique sequences PD FS, T2 FS and T1, as well as in coronal T1 sequences.
Adductor longus muscle-tendon injury to evaluate into the axial oblique sequences PD FS and T2 FS, as well as coronal STIR images.
Rectus abdominis tendinopathy to consider in sagittal STIR and axial oblique PD FS.
Rectus abdominis muscle-tendon injury to assess into the axial oblique sequences PD FS and T2 FS, as well in coronal STIR
Growth plate at pubic symphysis level to estimate in axial T1 sequences.

Furthermore, it was remembered the anatomical importance of the pre-aponeurotic fibrocartilaginous complex (PAFC). The PAFC is formed by the interconnection of the tendons of the adductor muscles and rectus femoris muscle and is included and integrated with the para-symphysarius ligaments and with the inguinal canal structures. Moreover, it is important to consider that the PAFC is in anatomical continuity with the symphysis central disc [59]. This complex anatomical structure represents a real anchoring central point and is therefore essentially formed by the interconnection of the fibres of the adductor muscles, the rectus abdominis, the external and internal oblique muscle, the inguinal ligament, the anterior pubic ligament, the arcuate ligament and the fibrocartilage symphysary disc. The acceptance of this anatomical concept presupposes two fundamental points: the first one is represented by the fact that the verification of the anatomical integrity PAFC is a central point of imaging exam and plays a crucial role in the formulation of the diagnosis, while the second point is the necessity to consider the "anatomical continuity" of the pubic symphysis, both of its superficial and deep anatomical structures and its functional continuity.

Conclusions

From the first GPS Italian consensus, some important points of discussion and reflection that we can summarize as follows emerged:

The controversy as regards the GPS diagnostic taxonomy can only be solved through the adoption of a common language, which satisfies the principles of clarity, fairness and sharability.
The adoption of a diagnostic pathway both from clinical point of view that concerning the imaging is a first step towards harmonizing and rationalizing the approach to

GPS. Obviously, such "guided" pathway does not limit the clinician professional skill, but rather it is a guide that would facilitate the formulation of definitive diagnosis, enabling this latter to be based on well-defined clinical diagnostic steps. Furthermore, the use of HAGOS questionnaire provides us the ability to objectively quantify the therapeutic effectiveness of the proposed procedures.

A standardized MRI protocol would facilitate the comparison of data from different study groups and substantially would favour the logical-deductive process that is the basis of the diagnostic path. In any case, it would require further and more detailed studies to clarify the true significance of some radiological findings that we can observe in a GPS framework.

Finally, the small number of female subjects observed in the studies present into the literature could theoretically be a limitation in the applicability of the data described above in a female population.

References

1. Bouvard M, Dorochenko P, Lanusse P, Duraffour H. La pubalgie du sportif—stratégie thérapeutique. J Traumatol Sport. 2004;21:146–63.
2. Ekberg O, Persson NH, Abrahamsson PA, et al. Longstanding groin pain in athletes: a multidisciplinary approach. Sports Med. 1988;6:56–61.
3. Weir A. From disruption to consensus: the thousand mile journey. Br J Sports Med. 2014;48:1075–7.
4. Serner A, van Eijck CH, Beumer BR, Hölmich P, Weir A, de Vos RJ. Study quality on groin injury management remains low: a systematic review on treatment of groin pain in athletes. Br J Sports Med. 2015;49(12):813.
5. Weir A, Brukner P, Delahunt E, Ekstrand J, Griffin D, Khan KM, Lovell G, Meyers WC, Muschaweck U, Orchard J, Paajanen H, Philippon M, Reboul G, Robinson P, Schache AG, Schilders E, Serner A, Silvers H, Thorborg K, Tyler T, Verrall G, de Vos RJ, Vuckovic Z, Hölmich P. Doha agreement meeting on terminology and definitions in groin pain in athletes. Br J Sports Med. 2015;49(12):768–74.
6. Vidalin H, Neouze G, Petit J, Brunet-Guedi E. Prise en charge chirurgicale des pubalgies du sportif. J Traumatol Sport. 2004;21:166–37.
7. Bisciotti GN, Eirale C, Vuckovic Z, Le Picard P, D'Hooge P, Chalabi H. La pubalgia dell'atleta: una revisione della letteratura. Med Sport. 2013;66(1):119–33.
8. Bisciotti GN, Auci A, Di Marzo F, Quaglia R, Volpi P. Groin pain syndrome: an association of different pathologies and a case presentation. Muscles Ligaments Tendons J. 2015;5(3):214–22.
9. Le Blanc KE, Le Blanc KA. Groin pain in athletes. Hernia. 2003;7:68–71.
10. Hölmich P. Long-standing groin pain in sportspeople falls into three primary patterns, a "clinical entity" approach: a prospective study of 207 patients. Br J Sports Med. 2007;41(4):247–52. discussion 252
11. Omar IM, Zoga AC, Kavanagh EC, Koulouris G, Bergin D, Gopez AG, Morrison WB, Meyers WC. Athletic pubalgia and "sports hernia": optimal MR imaging technique and findings. Radiographics. 2008;28(5):1415–38.
12. Jaap J. Longstanding adduction-related groin pain in athletes. Oisterwijk: Uitgeverij BOXpress; 2010.
13. Brophy RH, Prather H. Differential diagnosis of groin pain in athletes. In: Diduch DR, Brunt LM, editors. Sport hernia and athletic pubalgia. Diagnosis and treatment. London: Springer; 2014. p. 23–30.
14. Gilmore CJ, Diduch DR, Handley MV, Hanks JB. Sports hernia—History and physical examination: making the diagnosis with confidence. In: Diduch DR, Brunt LM, editors. Sport hernia and athletic pubalgia. Diagnosis and treatment. London: Springer; 2014. p. 23–30.
15. Lyons M, Brunt M. Sport hernia anatomy. What is a sport hernia? In: Diduch DR, Brunt LM, editors. Sport hernia and athletic pubalgia. Diagnosis and treatment. London: Springer; 2014. p. 1–11.
16. Sheen AJ, Stephenson BM, Lloyd DM, Robinson P, Fevre D, Paajanen H, de Beaux A, Kingsnorth A, Gilmore OJ, Bennett D, Maclennan I, O'Dwyer P, Sanders D, Kurzer M. Treatment of the sportsman's groin: British Hernia Society's 2014 position statement based on the Manchester Consensus Conference. Br J Sports Med. 2014;48(14):1079–87.
17. Sheen AJ, Jamdar S, Bhatti W. Calling for 'inguinal disruption' to be the term of choice for disorders of the inguinal ring: connecting Manchester and Doha. Br J Sports Med. 2015; doi:10.1136/bjsports-2015-095341. pii: bjsports-2015-095341 [Epub ahead of print].
18. Koulouris G. Imaging review of groin pain in elite athletes: an anatomic approach to imaging findings. AJR Am J Roentgenol. 2008;191:962–72.
19. Syme G, Wilson J, Mackenzie K, Macleod D. Groin pain in athletes. Lancet. 1999;353:1444.
20. Moeller JL. Sportsman's hernia. Curr Sports Med Rep. 2007;6:111–4.
21. Kachingwe AF, Grech S. Proposed algorithm for the management of athletes with athletic pubalgia (sports hernia): a case series. J Orthop Sports Phys Ther. 2008;38(12):768–81.
22. Gilmore J. Groin pain in the soccer athlete: fact, fiction and treatment. Clin Sports Med. 1998;17:787–93.
23. Meyers WC, Foley DP, Garrett WE, Lohnes JH, Mandlebaum BR. Management of severe lower abdominal or inguinal pain in high-performance athletes. PAIN (Performing Athletes with Abdominal or Inguinal Neuromuscular Pain Study Group). Am J Sports Med. 2000;28:2–8.
24. Ahumada LA, Ashruf S, Espinosa-de-losMonteros A, et al. Athletic pubalgia: definition and surgical treatment. Ann Plast Surg. 2005;55:393–6.

25. Diaco JF, Diaco DS, Lockhart L. Sports hernia. Oper Tech Sports Med. 2005;13:68–70.
26. Van Veen RN, de Baat P, Heijboer MP, et al. Successful endoscopic treatment of chronic groin pain in athletes. Surg Endosc. 2007;21:189–93.
27. Lynch SA, Renstrom PA. Groin injuries in sport: treatment strategies. Sports Med. 1999;28:137144.
28. Meyers WC, Lanfranco A, Castellanos A. Surgical management of chronic lower abdominal and groin pain in high-performance athletes. Curr Sports Med Rep. 2002;1:301–5.
29. Swan Jr KG, Wolcott M. The athletic hernia: a systematic review. Clin Orthop Relat Res. 2007;455:78–87.
30. Garvey JF, Read JW, Turner A. Sportsman hernia: what can we do? Hernia. 2010;14(1):17–25.
31. Hureibi KA, McLatchie GR. Groin pain in athletes. Scott Med J. 2010;55(2):8–11.
32. Hölmich P. Groin pain in football players. A systematic diagnostic approach. ASPETAR Sport Med J. 2013;2:192–6.
33. Hölmich P. Groin injuries in athletes—development of clinical entities, treatment, and prevention. Dan Med J. 2015;62(12):B5184.
34. Brown RA, Mascia A, Kinnear DG, Lacroix V, Feldman L, Mulder DS. An 18-year review of sports groin injuries in the elite hockey player: clinical presentation, new diagnostic imaging, treatment, and results. Clin J Sport Med. 2008;18(3):221–6.
35. Campanelli G. Pubic inguinal pain syndrome: the so-called sports hernia. Hernia. 2010;14(1):1–4.
36. Unverzagt CA, Schuemann T, Mathisen J. Differential diagnosis of a sports hernia in a high-school athlete. J Orthop Sports Phys Ther. 2008;38(2):63–70.
37. Kehlet H. Groin pain. Ugeskr Laeger. 2010;172(49):3393.
38. Preskitt JT. Sport hernia: the experience of Baylor University Medical Center at Dallas. Proc (Bayl Univ Med Cent). 2011;24(2):89–91.
39. Ross JR, Bedi A, Stone RM, Sibilsky Enselman E, Kelly BT, Larson CM. Characterization of symptomatic hip impingement in butterfly ice hockey goalies. Arthroscopy. 2015;31(4):635–42.
40. Lerebours F, Robertson W, Neri B, Schulz B, Youm T, Limpisvasti O. Prevalence of Cam-Type Morphology in Elite Ice Hockey Players. Am J Sports Med. 2016;44(4):1024–30. pii: 0363546515624671. [Epub ahead of print].
41. Bisciotti GN, Corradini B, Di Marzo F. La validazione del Copenhagen Hip and Groin Outcome Score (HAGOS) in lingua italiana nell'ambito del calcio. J Sport Traumatol. 2014;31(4):126–34.
42. Brennan D, O'Connell MJ, Ryan M, Cunningham P, Taylor D, Cronin C, O'Neill P, Eustace S. Secondary cleft sign as a marker of injury in athletes with groin pain: MR image appearance and interpretation. Radiology. 2005;235(1):162–7.
43. Lovell G, Galloway H, Hopkins W, Harvey A. Osteitis pubis and assessment of bone marrow edema at the pubic symphysis with MRI in an elite junior male soccer squad. Clin J Sport Med. 2006;16(2):117–22.
44. Cunningham PM, Brennan D, O'Connell M, MacMahon P, O'Neill P, Eustace S. Patterns of bone and soft-tissue injury at the symphysis pubis in soccer players: observations at MRI. AJR Am J Roentgenol. 2007;188(3):W291–6.
45. Zajick DC, Zoga AC, Omar IM, Meyers WC. Spectrum of MRI findings in clinical athletic pubalgia. Semin Musculoskelet Radiol. 2008;12(1):3–12.
46. Zoga AC, Kavanagh EC, Omar IM, Morrison WB, Koulouris G, Lopez H, Chaabra A, Domesek J, Meyers WC. Athletic pubalgia and the "sports hernia": MR imaging findings. Radiology. 2008;247(3):797–807.
47. Zoga AC, Mullens FE, Meyers WC. The spectrum of MR imaging in athletic pubalgia. Radiol Clin North Am. 2010;48(6):1179–97.
48. Balconi G. US in pubalgia. J Ultrasound. 2011;14:157–66.
49. Weir A, de Vos RJ, Moen M, Hölmich P, Tol JL. Prevalence of radiological signs of femoroacetabular impingement in patients presenting with long-standing adductor-related groin pain. Br J Sports Med. 2011;45(1):6–9.
50. Nepple JJ, Brophy RH, Matava MJ, Wright RW, Clohisy JC. Radiographic findings of femoroacetabular impingement in National Football League Combine athletes undergoing radiographs for previous hip or groin pain. Arthroscopy. 2012;28(10):1396–403.
51. Branci S, Thorborg K, Nielsen MB, Hölmich P. Radiological findings in symphyseal and adductor-related groin pain in athletes: a critical review of the literature. Br J Sports Med. 2013;47(10):611–9.
52. Powell JR., Nicholas CM., Viswanathan S.Anatomical and pictorial review of MRI findings in patients with athletica pubalgia—a trainees guide. EPOS™. 2013;C-1947:1–23.
53. Economopoulos KJ, Milewski MD, Hanks JB, Hart JM, Diduch DR. Radiographic evidence of femoroacetabular impingement in athletes with athletic pubalgia. Sports Health. 2014;6(2):171–7.
54. Branci S, Thorborg K, Bech BH, Boesen M, Nielsen MB, Hölmich P. MRI findings in soccer players with long-standing adductor-related groin pain and asymptomatic controls. Br J Sports Med. 2015;49(10):681–91.
55. Lee RK, Griffith JF, Ng WH. High accuracy of ultrasound in diagnosing the presence and type of groin hernia. J Clin Ultrasound. 2015;43(9):538–47.
56. Miller J, Cho J, Michael MJ, Saouaf R, Towfigh S. Role of imaging in the diagnosis of occult hernias. JAMA Surg. 2014;149(10):1077–80.
57. Robinson P, Grainger AJ, Hensor EM, Batt ME, O'Connor PJ. Do MRI and ultrasound of the anterior pelvis correlate with, or predict, young football players' clinical findings? A 4-year prospective study of elite academy soccer players. Br J Sports Med. 2015;49(3):176–82.
58. Morley N, Grant T, Blount K, Omar I. Sonographic evaluation of athletic pubalgia. Skeletal Radiol. 2016 45(5):689-699. [Epub ahead of print].
59. MacMahon PJ, Hogan BA, Shelly MJ, Eustace SJ, Kavanagh EC. Imaging of groin pain. Magn Reson Imaging Clin N Am. 2009;17(4):655–66.

2 Inguinal Region Anatomy

Umile Giuseppe Longo, Vincenzo Candela, Giuseppe Salvatore, Mauro Ciuffreda, Alessandra Berton, and Vincenzo Denaro

2.1 Introduction

Groin pain is typically a multifactorial condition characterized by vague and diffuse pain extended between the lower abdomen and medial thigh. It may be attributed to a variety of diagnostic entities and requires a differential diagnosis [1–4].

To understand groin pain pathogenesis, a precise anatomical knowledge of the pubic region is required.

Even though there is not a consensus on the anatomic definition of the groin, it can be defined as the region extended from the distal aspect of the abdominal wall to the proximal adductor compartment of the thigh, encompassing pubic symphysis [5].

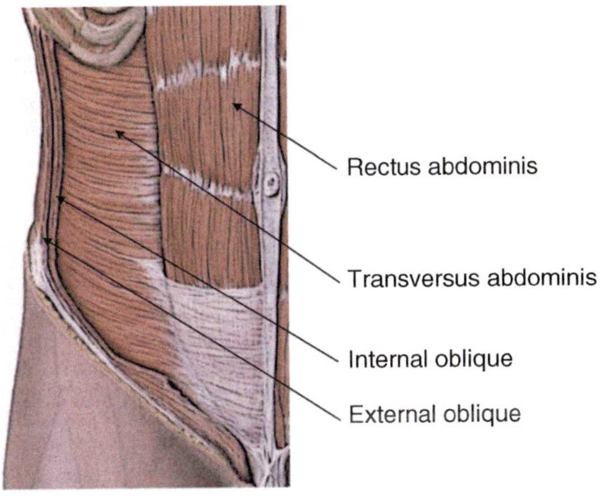

Fig. 2.1 Abdominal wall

2.2 The Abdominal Wall

The abdominal wall (Fig. 2.1) is formed by nine overlying layers that from the surface to the depth are the skin, the subcutaneous tissue, the superficial fascia, the external oblique muscle, the internal oblique muscle, the transversus abdominis muscle, the transversalis fascia, the preperitoneal adipose and areolar tissue and the peritoneum [6].

The superficial fascia, also known as Scarpa's fascia, is a dense fibrous connective layer contiguous with the fascia lata of the thigh.

The external oblique muscle originates from the lower seven ribs, from the thoracolumbar sheath, from the outer lip of the iliac crest and from the inguinal ligament. The muscle, anteriorly, near the midclavicular line, become a strong aponeurosis that passes anteriorly to the rectus muscle to insert into the linea alba.

The inferior edge of the external oblique muscle aponeurosis forms the inguinal ligament (Poupart's ligament) that is extended from the anterior superior iliac spine to the pubic tubercle [7].

U.G. Longo, M.D., M.Sc., Ph.D. (✉) • V. Candela • G. Salvatore • M. Ciuffreda • A. Berton • V. Denaro
Department of Orthopaedic and Trauma Surgery, Campus Bio-Medico University, Via Alvaro del Portillo, 200, Trigoria, Rome 00128, Italy
e-mail: ug.longo@gmail.com

R. Zini et al. (eds.), *Groin Pain Syndrome*, DOI 10.1007/978-3-319-41624-3_2

The internal oblique muscle arises from the lower five ribs, from the thoracolumbar sheath, from the intermediate lip of the iliac crest and from the lateral half of the inguinal ligament. The muscle, anteriorly, become a strong aponeurosis. Above the line of Douglas, this aponeurosis is divided into anterior and posterior sheaths that pass anteriorly and posteriorly to the rectus abdominis, respectively. Below the line of Douglas, instead, the aponeurosis passes anteriorly to the rectus muscle. The lower fibres of the internal oblique muscle insert between the symphysis pubis and pubic tubercle. Some fibres, furthermore, form the cremasteric muscle.

The transversus abdominis muscle originates from the lower five ribs, from the thoracolumbar sheath, from the inner lip of the iliac crest and from the lateral half of the inguinal ligament. Anteriorly, the muscle becomes an aponeurotic sheet that passes posteriorly to the rectus abdominis above the line of Douglas and anteriorly to the rectus muscle below the line.

The aponeurosis of the anterolateral muscles, anteriorly, melt around the rectus abdominis muscle and formed the rectus sheath.

The transversalis fascia contributes to the structural integrity of the abdominal wall covering the deep surface of the transversus abdominis muscle.

The rectus abdominis muscles act as the major abdominal wall stabilizer. They originate from the anterior surface of the fifth, sixth and seventh costal cartilages and from the xiphoidal process. Their insertions are on the superior aspect of the pubic crest just lateral to the pubic symphysis, and they are connected near the anterior midline by the linea alba.

The rectus sheath is reported to be continuous with adductor longus via the pubic symphysis capsular tissues. This confluence of soft tissue structures anterior to the pubic symphysis may provide the anatomical substrate for a stabilizing or force transmission mechanism [8–10].

Abdominal wall haematic supply comes from the last six intercostal arteries, four lumbar arteries, superior and inferior epigastric arteries and deep circumflex iliac arteries.

2.3 The Inguinal Canal

The inguinal canal is an about 4 cm canal extended between the internal and the external inguinal rings. It contains the spermatic cord (formed by cremasteric muscle fibres, testicular artery and vein, genital branch of the genitofemoral nerve, vas deferens, cremasteric vessels, lymphatic and processus vaginalis) in men and the round ligament in woman.

The superficial wall of the canal is formed by the external oblique aponeurosis, the cephalad wall by the internal oblique and transversus abdominis aponeurosis, the inferior wall by the inguinal ligament and the lacunar ligament and the posterior wall by the transversus abdominis muscle and transversalis fascia.

The groin area has important sensory nerves: iliohypogastric, ilioinguinal nerves and genital branch of genitofemoral nerve.

2.4 The Pubic Symphysis

The pubic symphysis is an amphiarthrodial joint.

It connects the two pubic bones via a fibrocartilaginous articular disc, and it has no joint capsule. The joint is supported anteriorly by the anterior pubic ligament, inferiorly by the arcuate ligament and superiorly by the pubic ligament.

The pubic symphysis dissipates the heavy forces from the lower limbs and allows minimal movements [11].

2.5 The Hip

The hip (Figs. 2.2 and 2.3) is an enarthrodial joint formed by the articular surface of the femoral head and by the cavity of the acetabulum. The femoral head is covered by the articular cartilage with the exception of the fovea capitis femoris, point of origin of the ligamentum teres. The acetabulum, instead, has an incomplete marginal ring of cartilage called lunate surface. Furthermore, the acetabulum has a central depression occupied by fat covered by synovial membrane [12].

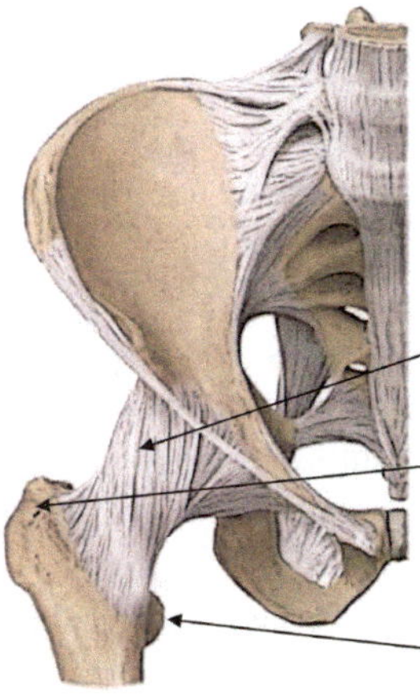

Fig. 2.2 Hip AP

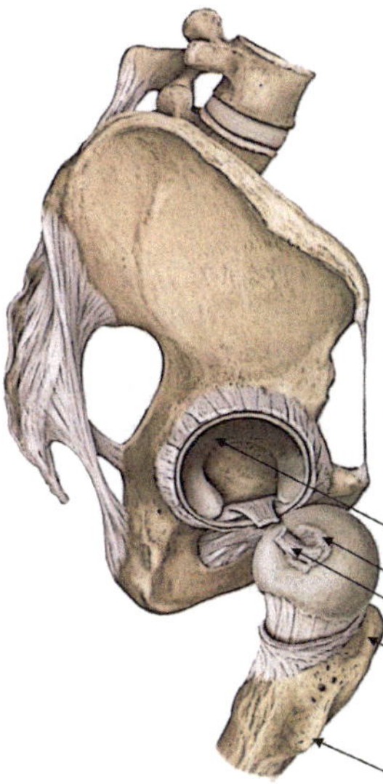

Fig. 2.3 Adductor compartment

The ligamentum teres is implanted on the acetabular notch, and it is tense when the thigh is semiflexed and the limb adducted or rotated outwards and relaxed when the limb is abducted.

The joint is supported by the articular capsule, by the iliofemoral ligament, by the ischiofemoral ligament, by the pubofemoral ligament, by the ligamentum teres femoris and by the glenoid labrum.

The articular capsule is composed by circular fibres in the deep area and by longitudinal fibres in the superficial area. The longitudinal fibres are greatest in amount at the upper and front part of the capsule, and they are reinforced by capsular ligaments. The capsule is attached to the margin of the acetabulum, to the outer margin of the labrum and to the transverse ligament proximally. Distally, it is attached to the intertrochanteric line anteriorly, to the neck above the intertrochanteric crest posteriorly and to the lower part of the neck inferiorly, near the lesser trochanter.

The iliofemoral ligament, called also Y-ligament or ligament of Bigelow, originates between the lower part of the anterior inferior iliac spine and the acetabular margin; it divides into two bands, one of which is fixed to the lower part of the intertrochanteric line and the other to the upper part of the same line.

The ischiofemoral ligament is sited posteriorly, springs from the ischiatic rim of the acetabulum and blends with the capsular fibres to insert in the posterior area of the femoral neck. It controls the internal rotation and the adduction when the hip is flexed.

The pubofemoral ligament originates from the obturator crest and from the superior ramus of the pubis to insert near the lesser trochanter; its fibres blend with the capsule and with the deep surface of ischiofemoral ligament.

The glenoid labrum is a fibrocartilaginous semicircular rim sited at the acetabular margin and competed at the bottom by the transverse ligament. It closely surrounds the head of the femur.

Finally, the synovial membrane originates from the glenoid labrum and inserts at the margin of the cartilaginous surface of the femoral head [13–16].

2.6 The Adductor Compartment

The adductor compartment (Fig. 2.4) is involved in the stability of the anterior pelvis and pubic joint.

The gracilis is the most superficial muscle on the medial side of the thigh. It originates from the upper part of the pubic arch and from the anterior margins of the lower part of the pubis symphysis. The muscle fibres run vertically and pass behind

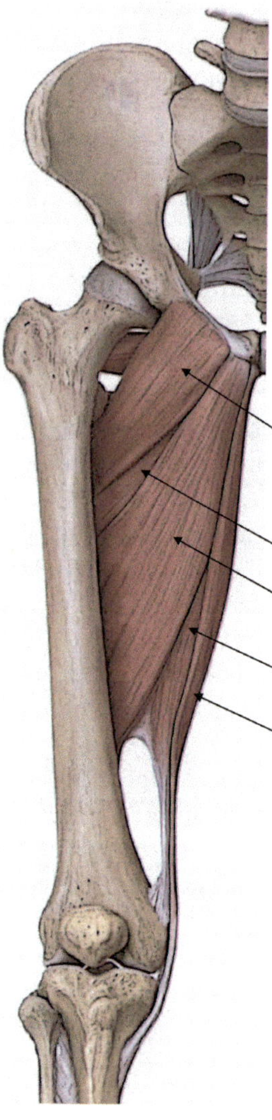

Fig. 2.4 Hip LL

the medial femoral condyle. The insertion is on the upper part of the medial surface of the tibia. Semitendinosus, gracilis and sartorius have a common insertion into the anterior-medial aspect of the tibia called the pes anserinus.

The pectineus is a quadrangular muscle that originates from the pectineal line of the pubis and inserts into the pectineal line of the femur.

The adductor longus is a superficial adductor muscle that arises from the body of the pubis in the angle between the crest and the symphysis. Its tendon is thin anteriorly and composed of muscular fibres on the deep surface of the pubic attachment.

The adductor longus inserts by an aponeurosis into the linea aspera, between the vastus medialis and the adductor magnus.

The adductor brevis is a triangular muscle that arises from the outer surfaces of the superior and inferior rami of the pubis, between the gracilis and obturator externus, and inserts by an aponeurosis into the line leading from the lesser trochanter to the linea aspera and into the upper site of the linea aspera, behind the pectineus and the adductor longus.

The adductor magnus is a large triangular muscle that originates from the inferior pubic ramus and ischial tuberosity, while the remainder arise from the anterior aspect of the superior pubic ramus and the pubic tubercle. His insertion is on the femoral linea aspera [10].

References

1. Orchard J, Wood T, Seward H, Broad A. Comparison of injuries in elite senior and junior Australian football. J Sci Med Sport. 1998;1(2):83–8.
2. Besjakov J, von Scheele C, Ekberg O, Gentz CF, Westlin NE. Grading scale of radiographic findings in the pubic bone and symphysis in athletes. Acta Radiol. 2003;44(1):79–83.
3. Maffulli N, Loppini M, Longo UG, Denaro V. Bilateral mini-invasive adductor tenotomy for the management of chronic unilateral adductor longus tendinopathy in athletes. Am J Sports Med. 2012;40(8):1880–6.
4. Martinelli N, Longo UG, Marinozzi A, Franceschetti E, Costa V, Denaro V. Cross-cultural adaptation and validation with reliability, validity, and responsiveness of the Italian version of the Oxford Hip Score in patients with hip osteoarthritis. Qual Life Res. 2011;20(6):923–9.
5. Koulouris G. Imaging review of groin pain in elite athletes: an anatomic approach to imaging findings. AJR Am J Roentgenol. 2008;191(4):962–72.
6. Condon RE. Reassessment of groin anatomy during the evolution of preperitoneal hernia repair. Am J Surg. 1996;172(1):5–8.
7. Punwar S, Khan WS, Longo UG. The use of computer navigation in hip arthroplasty: literature review and evidence today. Ortop Traumatol Rehabil. 2011;13(5):431–8.
8. Robertson BA, Barker PJ, Fahrer M, Schache AG. The anatomy of the pubic region revisited: implications for the pathogenesis and clinical management of chronic groin pain in athletes. Sports Med. 2009;39(3):225–34.
9. Lovell G, Galloway H, Hopkins W, Harvey A. Osteitis pubis and assessment of bone marrow edema at the

pubic symphysis with MRI in an elite junior male soccer squad. Clin J Sport Med. 2006;16(2):117–22.
10. Standring S. Gray's anatomy: the anatomical basis of clinical practice. New York: Elsevier; 2005.
11. Gamble JG, Simmons SC, Freedman M. The symphysis pubis. Anatomic and pathologic considerations. Clin Orthop Relat Res. 1986;203:261–72.
12. Zini R, Longo UG, de Benedetto M, Loppini M, Carraro A, Maffulli N, et al. Arthroscopic management of primary synovial chondromatosis of the hip. Arthroscopy. 2013;29(3):420–6.
13. Byrd JW, Pappas JN, Pedley MJ. Hip arthroscopy: an anatomic study of portal placement and relationship to the extra-articular structures. Arthroscopy. 1995;11(4):418–23.
14. Toogood PA, Skalak A, Cooperman DR. Proximal femoral anatomy in the normal human population. Clin Orthop Relat Res. 2009;467(4):876–85.
15. Martin HD, Savage A, Braly BA, Palmer IJ, Beall DP, Kelly B. The function of the hip capsular ligaments: a quantitative report. Arthroscopy. 2008;24(2):188–95.
16. Longo UG, Franceschetti E, Maffulli N, Denaro V. Hip arthroscopy: state of the art. Br Med Bull. 2010;96:131–57.

3 Physiology and Biomechanics

Papalia Rocco, Vasta Sebastiano, Torre Guglielmo, Ciuffreda Mauro, and Denaro Vincenzo

3.1 Introduction

The groin is one of the most complex regions of the human body, due to several osteo-muscular and visceral structures that merge into this area and participate in complex functions, especially concerning movement. In this anatomic area, the muscle of the trunk and lower extremities and osteo-fibrous structure (iliac spine and pubis, inguinal ligaments and fascia) converge and surround the hip, making this joint very complicated anatomically. In the groin, the ileo-inguinal region is the most stressed area, especially in the anterior and lateral flexion, due to the insertion, in the upper region, of the large enthesis of the abdominal muscles and, in the lower region on the medial part of the pubis, of adductor and pectineus muscles [1]. The genesis of the groin pain can be triggered by several factors, making, some time, it very difficult to understand the cause of illness of patients. The groin pain is often caused by hip arthritis and morphological abnormalities such as femoro-acetabular impingement (FAI), which is very common in the athletic population, although muscular pathologies including tendinitis, enthesitis and muscular strain can participate to the genesis of pain [2]. Actually, it has been demonstrated, with a strong evidence [3], that muscle function is the most impaired when groin pain syndrome occurs, with significant differences in muscular testing outcomes between healthy and affected subjects. At the same time, the groin pain can be caused by visceral components that cross the inguinal canal: the spermatic cord in the male, the round ligament of the uterus in the female and especially in case of the visceral herniation. In this chapter, an overview will be provided on normal physiology and biomechanics of the inguinal region and hip joint, paying particular attention to the elements that are mainly involved in the genesis of groin pain.

3.2 Hip Joint Function

The hip joint is an enarthrosis. The acetabulum with a concave surface and femoral head with a convex spherical one constitutes the joint. This type of joint allows movement on each of the spatial planes, including axial rotation. The cartilaginous acetabular labrum and capsular and extra-capsular ligaments that surround the femoral head are classified as static stabilizers. Instead, the musculoligamentous structures are classified as dynamic ones [4]. The balance that is created between these complex structures is one of the key areas for whole body static and dynamic

P. Rocco, M.D., M.Sc., Ph.D. (✉) • V. Sebastiano
T. Guglielmo • C. Mauro • D. Vincenzo
Department of Orthopaedic and Trauma Surgery, Campus Bio-Medico University of Rome, Via Alvaro del Portillo, 200, Trigoria, Rome 00128, Italy
e-mail: r.papalia@unicampus.it

R. Zini et al. (eds.), *Groin Pain Syndrome*, DOI 10.1007/978-3-319-41624-3_3

balance; the correct function of the hip joint and its stabilization allow the upright and the gait.

3.3 Physiology of the Hip

The articular cartilage of the hip has a normal width of 2 mm in healthy adult subjects that encounters a drastic thinning when osteoarthritic process begins. The width of an arthritic hip cartilage is near to 0 mm, yielding to an uncovering of the subchondral bone and determining the arthritic pain during the load. Articular cartilage biophysic behaviour can be compared to a biphasic, linear and porous material. Synovial fluid has a key role in the homeostasis of joint space viscosity. It can be assimilated to a Newtonian fluid, which viscosity is inversely proportional to the shear stress. Thus, synovial fluid lacks or decreases during arthritic process and imposes the need for viscosupplementation therapy with hyaluronic acid or similar composites. Femoral epiphysis is the exact point where the force vectors converge. Therefore, this area requires a strong structure, which is provided by the great and small trochanter that enlarge and reinforce the bone shape and, at the same time, allow several muscular insertions. The femoral head and neck have a thin cortical bone layer that enriches and becomes thicker in the epiphyseal region. The Wolff law [5] describes the way of bone remodelling that follows static and dynamic loading, taking the femoral epiphysis as the main model. It affirms that bone remodelling follows the loading curve, with the trabecular bone that is disposed in the direction of the stress force applied to the area.

3.4 Biomechanics and Movement

The hip joint has three degrees of freedom into the three spatial axes: the sagittal axis, the vertical and the anteroposterior one; therefore, the movements admitted are internal rotation, external rotation (vertical axis), flexo-extension (sagittal axis) and adduction-abduction (anteroposterior axis). Thus, the range of movement of the hip is really wide, allowing gait, running and other main activities of the human body [6]. In the following paragraphs, the single movements are described in details:

Flexion—It is mainly influenced by the position of the knee, since a flexed knee allows a hip flexion of 120°, while knee extension allows a 90° hip flexion. Muscles that are mainly involved in the flexion movements are iliopsoas, sartorius, anterior rectus and tensor of the fascia.

Extension—This movement is influenced by knee too. The knee flexion provides a hip extension of 10°, which became 20° with knee extension. At the same time, extension of the hip is limited by several anterior structures including ileo-femoral ligament, anterior capsule fibres and flexor muscles. The hip extension is allowed by gluteal muscles, hamstrings and femoral biceps. The insertion of greater, medium and smaller gluteus on the proximal femoral epiphysis provides also mechanical stability and proprioception of the hip.

Abduction—A 45° abduction is allowed, with a specific counterbalancing of the lumbar spine to face lateral tilt of the pelvis. Muscles involved in abduction movement include the gluteal, piriformis and tensor of the fascia.

Adduction—Adduction is limited to 30°. Physiologically, adduction is always paired with leg flexion or extension. Muscles involve in this movement are the great adductor, adductor brevis, adductor longus, hamstrings, femoral biceps, internal and external obturators and pectineus.

Internal rotation—Internal rotation is about 60° and involves the tensor of the fascia, anterior fascicles of medium gluteus and small gluteus.

External rotation—This movement is about 30° and muscles involved are internal and external obturators, piriformis, pectineus, quadratus, greater gluteus and posterior fascicles of the medium gluteus.

The entire movement provided by the single movements described before is defined "circumduction" of the hip, with a significant variation of the hip centre [7].

The complete hip movement assumes a relevant role in daily activities, such as sitting and standing up, ascending and descending stairs and wearing shoes and socks.

3.5 Force Loading of the Hip

During activities, the joint is constantly subjected to loading. The higher peak forces are reached during the deambulation, instead during standing position, 80% of body weight force load on the joint. In the erect position, weight is transmitted from the fifth lumbar vertebra, where a first force breakdown occurs, to the base of the sacrum and thus to the sacroiliac joint bilaterally. The acetabulum and femoral head receive then this loading as a 50% of the upper body weight. A two-dimensional model, called Pauwel's model [8], has been introduced to explain the hip loading. It assumes frontal bi-dimensional plane view of a static condition, with a single leg stance position. Joint reaction force (JRF) is defined as that force that the femoral head exerts on the acetabulum. A vector sum can be therefore plotted (Fig. 3.1), where the JRF counterbalances the sum of the body weight force (*W*) and abductor muscle force (*A*). Since a static model is proposed, the sum of the force moments is null. Therefore, abductor force can be calculated (Eq. 3.1) as a function of the body weight force (5/6 of the total weight, because the substanding leg weight is not obviously loaded on the hip), its arm (*d*) and the arm of abductor force itself (*l*), where, *d* is equal to the distance between the centre of the pelvis and the femoral head centre and *l* is equal to the distance between the centre of the femoral head and the greater trochanter tip:

$$\vec{A} = \frac{5/6\vec{W} \times d}{l} \tag{3.1}$$

Thus, knowing the body weight of the individual and his abductor force, the resulting force (JRF) can be calculated as a vector force (Eq. 3.2):

$$\overrightarrow{JRF} = -\left(\vec{W} + \vec{A}\right) \tag{3.2}$$

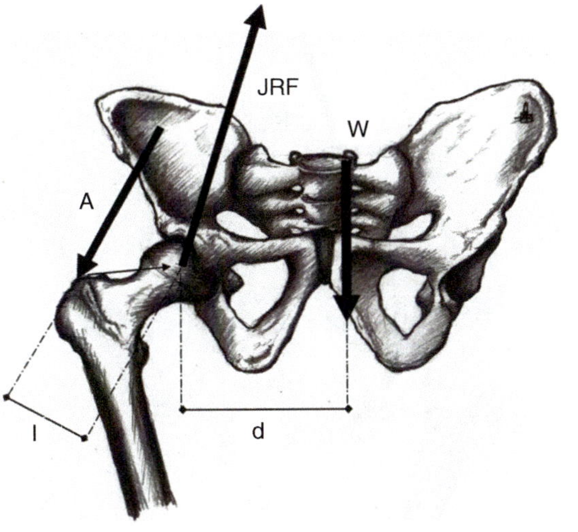

Fig. 3.1 Static model of vector forces exerting on the hip. *A* is the force of abductor muscles, *W* is the body weight and *l* and *d* represent their arms. *JRF* is the Joint Reaction Force whch counterbalances the aforementioned forces on the femoral head

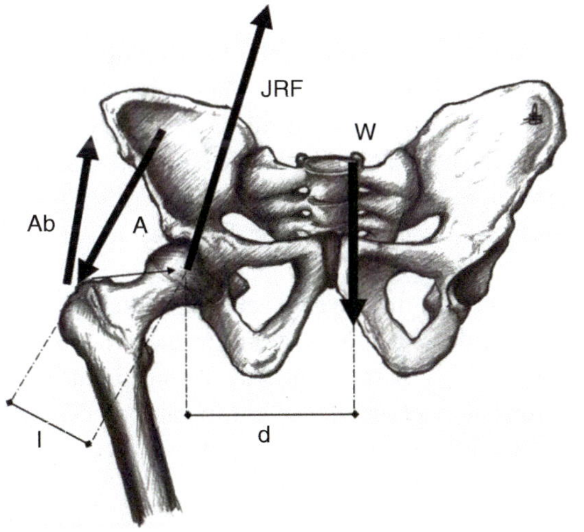

Fig. 3.2 Abductor force (*Ab*) represents the reaction of gluteal muscles to single leg stance position

It usually assumes a value of about 2.7-fold the body weight and the vector has a vertical inclination of about 18°.

Pauwel's model can be easily applied to the clinical practice, since single leg stance position is clearly reproduced during slow gait or asking to the patient to maintain single leg stance.

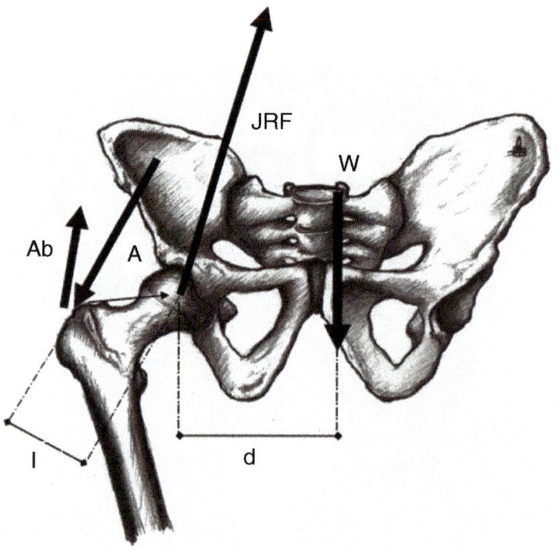

Fig. 3.3 When gluteal insufficiency occurs and abductor reaction (*Ab*) decreases, *JRF* increases, as *W* force is closer to rotational center of the femoral head (*d* decreses)

Trendelenburg sign is the most important clinical sign to evaluate abductor muscle force, since in case of abductor insufficiency, the patient cannot maintain a correct horizontal balance of the pelvis during single leg stance. Therefore, when this occurs, the weight is moved more closely to the rotational centre of the femoral head, providing a shortening of the body weight force arm (*d*) and consequently increasing the joint reaction force (Figs. 3.2 and 3.3).

Therefore, the use of crutches during gait helps to decrease the loading on the hip (so the JRF) by 50%. In the same way, losing about 20% of the body weight can be comparable to the use of crutches.

3.6 Hip Function During Gait

Gait process modifies continuously the hip position, with a maximum extension during heel-off phase and maximum flexion during heel-strike phase. Muscles function varies during these phases [9]. In the heel-strike phase, hip extension begins through the activation of hamstring and greater gluteus muscles. Subsequently, at the beginning of the heel-off phase, the tensor of the fascia and medium and small gluteus provide complete extension with anterior propulsion of the body. Heel-off phase is characterized by the activation of iliac muscles, and in toe-off phase, the adductor longus is also involved and participates in providing load shift to the contralateral limb. In the swing phase, the group of muscles activated to allow the hip flexion and anterior propulsion of the limb are the iliopsoas, gracilis, femoral rectus and sartorius. Eventually, hamstrings and greater gluteus activate to prepare to the following stance phase [10].

Considering the aforementioned modifications of muscular function, a biphasic two-peak pattern of force loading is provided to the femoroacetabular joint. The first one occurs during heel-strike phase and the second one during heel-off one. During this variation of the load, the cartilage is deformed to increase the contact area between articular surfaces, decreasing synovial fluid lateral flow speed and increasing time of action of the squeeze film. The squeeze film is the normal interface between two surfaces, separated by a fluid. In this case, synovial fluid is interposed between the femoral and acetabular cartilage and provides a squeeze film whose time of action can reach 10 s when maximum loading is applied. The squeeze film mechanism occurs during heel-off phase, when the fluid is compressed between articular surfaces. Conversely, in the other gait phases, joint space widens; therefore, the synovial fluid flows into the joint cavity.

In sports medicine, a relevant role is covered by the analysis of dynamic forces loading on the hip, during gait and running, but also during the several movements that the different sports activities are characterized. It is actually possible to calculate the whole pattern of applied loads, although these computations can be carried out only knowing the complete kinetic and kinematics of the inferior limb. Due to this complexity, this calculation is beyond the aim of the present chapter. However, an approximate estimation of these loading forces is proposed. It should be considered that during normal gait, a three- to fourfold body weight force is applied to the hip that increases to a sevenfold force during running, with a maximum load during heel-strike phase. These forces increase even more in specific sports activities such as American football, tennis and basketball, where a high number of torsions and rotational movements are required. For example, Van der Bogert et al. reported a force analysis on freestyle skiers where a 8.3–12.4-fold body weight force is applied to the hip during activity.

References

1. Callaghan JJ, Rosemberg AG, Rubash HE. The adult hip. Baltimore: Lippincott Williams & Wilkins; 2007.
2. Bradshaw CJ, Bundy M, Falvey E. The diagnosis of longstanding groin pain: a prospective clinical cohort study. Br J Sports Med. 2008;42(10):851–4. doi:10.1136/bjsm.2007.039685.
3. Mosler AB, Agricola R, Weir A, Holmich P, Crossley KM. Which factors differentiate athletes with hip/groin pain from those without? A systematic review with meta-analysis. Br J Sports Med. 2015;49(12):810. doi:10.1136/bjsports-2015-094602.
4. Zaffagnini S, Signorelli C, Bonanzinga T, Lopomo N, Raggi F, Di Sarsina TR, Grassi A, Marcheggiani Muccioli GM, Marcacci M. Soft tissues contribution to hip joint kinematics and biomechanics. Hip Int. 2016;26(Suppl 1):23–7. doi:10.5301/hipint.5000407.
5. Boyle C, Kim IY. Three-dimensional micro-level computational study of Wolff's law via trabecular bone remodeling in the human proximal femur using design space topology optimization. J Biomech. 2011; 44(5):935–42. doi:10.1016/j.jbiomech.2010.11.029.
6. Bonneau N, Bouhallier J, Haeusler M, Gagey O. An innovative method for the three-dimensional study of hip joint motion. Comput Methods Biomech Biomed Engin. 2014;17(Suppl 1):56–7. doi:10.1080/10255842.2014.931110.
7. Kainz H, Carty CP, Modenese L, Boyd RN, Lloyd DG. Estimation of the hip joint centre in human motion analysis: a systematic review. Clin Biomech. 2015;30(4): 319–29. doi:10.1016/j.clinbiomech.2015.02.005.
8. Verbrugge J. Apropos of the report of Prof. Pauwels. General summary of the treatment of deforming arthroses of the hip. Acta Chir Belg. 1964;63:77–80.
9. Bergmann G, Deuretzbacher G, Heller M, Graichen F, Rohlmann A, Strauss J, Duda GN. Hip contact forces and gait patterns from routine activities. J Biomech. 2001;34(7):859–71.
10. Quesnel T, Gueritey PM, Gonon GP. Biomechanics of the hip: forces exerted during walking. Surg Radiol Anat. 1995;17(3):249–53.

4 Radiologic Diagnosis

Alessio Auci

4.1 Imaging Techniques

GPS can be subdivided into three main categories, based both on the etiopathogenesis and on the timing of onset/disappearance of the clinical framework, namely:

1. The GPS of **traumatic origin**: the onset of pain is due to a precise traumatic event, and this hypothesis is supported by the anamnestic investigation, by clinical examination, and by imaging.
2. The GPS due to **functional overload**: characterized by insidious and progressive onset, the patient has no memory of trauma or a situation to which is attributed with certainty the onset of pain symptoms.
3. The **long-standing GPS** (LSGPS) or **chronic GPS**: the cohort of symptoms complained by the patient continues for a long period and is recalcitrant to any conservative therapy.

The most frequent *acute injuries* in the pelvic, inguinal, and hip region are groin injuries and labral tears. Acute groin injuries affect mostly the adductors, hip flexor muscles, and lower abdominal and oblique muscle aponeuroses. In the skeletal immature younger athlete, traction injury at the apophysis (growth plate) may lead to pain and dysfunction, apophysitis, or even avulsion fractures. Stress fracture, secondary to increased training load, may cause acute onset groin pain. These injuries are often accompanied by other intra-articular pathologies such as labral tears, torn ligamentum teres, torn capsule, and chondral injuries.

Insidious and painful *chronic injuries* in the pelvic, inguinal, and hip region are among the most difficult injuries to deal with, both with respect to diagnosis and treatment; therefore, these injuries represent "the big diagnostic challenge." The most common pain generators are insertional tendinopathy of the adductor longus, rectus femoris, rectus abdominis, and iliopsoas. Osteitis pubis is generally due to overuse and overload in athletes with imbalanced muscular forces around the pelvis. Hip disorders, in particular femoroacetabular impingement (FAI), can predispose to subsequent chondrolabral lesions and may cause progression of osteoarthritis in the hip.

Radiographs (RX) should be commonly the first diagnostic studies needed for the evaluation of sports skeletal injuries. RX is requested, usually after a trauma, to detect or exclude fractures or apophysitis/avulsion fractures in the skeletal immature players. In the insidious or chronic onset of symptoms, RX may also show abnormal bony morphology of the hip joint (e.g., femoroacetabular impingement (FAI)) and degenerative changes of the symphysis pubis such as

A. Auci
Kinemove Rehabilitation Center,
Via Veterani dello Sport, 1, Pontremoli 54027, Italy
e-mail: alessioauci@gmail.com

R. Zini et al. (eds.), *Groin Pain Syndrome*, DOI 10.1007/978-3-319-41624-3_4

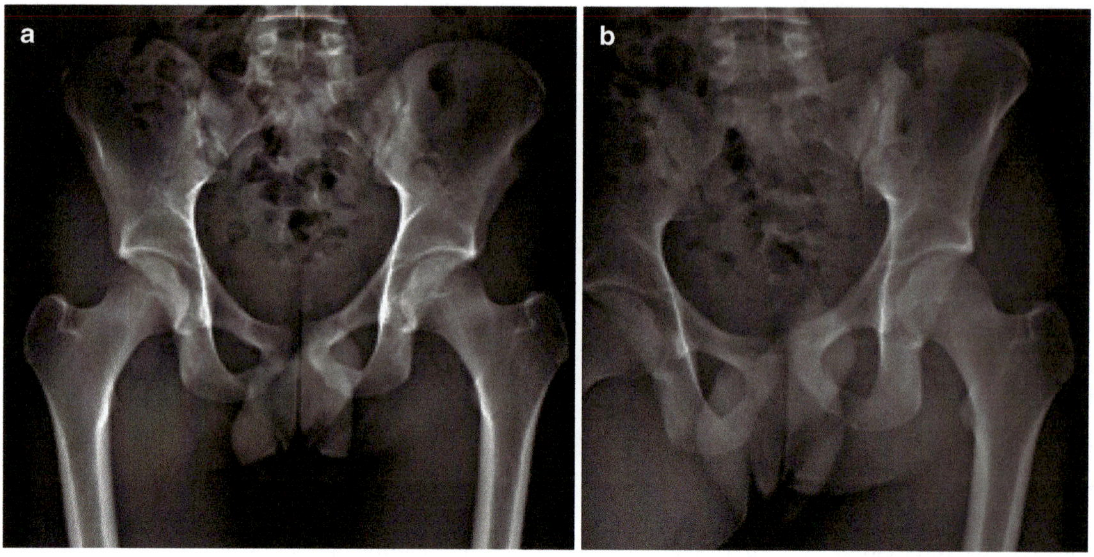

Fig. 4.1 RX AP flamingo view demonstrating (**b** image) a 3 mm left-upward symphysis excursion, indicating pubic instability in standard anterior posterior view (**a** image) and with monopodalic alternate bearing (**b** image)

subchondral sclerosis, symphyseal irregularity, and bone resorption pathology (e.g., signs of chronic pelvic instability), detect soft tissue calcification (e.g., at the adductor, rectus femoris origin, or iliopsoas insertion), or determine risk factor of leg length discrepancy [1]. The typically radiologic projections performed in **acute injuries** should be taken initially in the standard anteroposterior (AP) and lateral views. If a **chronic injury** is suspected (e.g., femoroacetabular impingement or mechanical instability of the symphysis pubic), the following tailored RX imaging protocol is suggested: upright pelvic anteroposterior (AP) view, 45° or 90° Dunn's view, and flamingo stress radiographs (e.g., anterior posterior view in upright position with monopodalic alternate bearing) (Fig. 4.1).

An alpha angle greater than 50° measured on the Dunn projection or greater than 69° on the standard AP view establishes a radiologic diagnosis of cam FAI. Pubic instability is directly measured as the amount of vertical displacement (vertical offset greater than 2 mm) or widening (greater than 7 mm) observed at the symphysis on "flamingo" stress radiographs [2–4].

Ultrasound (US) is a valuable tool in focused assessment of both acute and chronic groin pain and allows for dynamic evaluation and, of course, is also very useful in image-guided diagnostic intervention. High-frequency linear array probes are used to perform musculoskeletal US examinations; during the examination, it is of paramount relevance to palpate with the probe on the skin the point of maximal tenderness [5]. To avoid artifacts or pitfalls, comparison with the opposite side is required. US is used in the initial assessment and diagnostic evaluation of tendons and muscle injuries including para-symphyseal tendinopathies or tears at one or more location, tendon snapping, or joint fluid in the younger athletes. In low-grade or chronic injuries of the adductor and hamstrings, US has a low sensitivity for detecting tenoperiosteal-entheseal disease. Especially the hamstring enthesis is difficult to assess with US because of its deep location. The capability of US real time imaging and the possibility to perform provocative stress maneuvers during the US scanning protocols, consent US-guided focused clinical examination; US therefore is a useful technique in integrated imaging protocol for depiction of clinically occult groin hernias or weakness of the posterior inguinal wall (e.g., sports hernia) [6].

Computed tomography (CT) imaging is a helpful imaging tool for the evaluation of bony abnormalities in athletes requiring surgical intervention (e.g., arthroscopic surgery); from the data set, images can be reformatted in other planes (2-D multiplanar technique) or be used for volume rendering (3-D technique) for a more effective display of complex anatomic and pathologic structures. Furthermore, collision detection algorithm when using 3-D models from CT scan can localize regions of anticipated mechanical impingement in symptomatic patients with hip pain requiring surgery [7].

Magnetic resonance imaging (MRI) of the pelvis as well as imaging of the groin and symphyseal region has proven to be the "gold standard" imaging modality in diagnosis of acute and chronic soft tissue injuries as well as radiographically occult and subtle fractures. The advantages of MRI include its multiplanar capabilities, excellent soft tissue and bone contrast, and lack of ionizing radiation. A closed-bore, high-field system (1.5 T or greater) will generate the best images. While a reasonably small field of view and high imaging matrix are desirable to facilitate higher spatial resolution (e.g., tendon attachments to the pubis), MRI protocol should include at least a screening examination of the entire pelvis to rule out visceral causes of groin pain. Musculoskeletal MRI commonly includes a combination of pulse sequences in both long-axis (sagittal or coronal) and short-axis (axial) planes. Axial oblique imaging perpendicular to the pubic bodies is a useful tool for depicting the tendon attachments and better separating the adductor muscles from each other. Short TR/TE (T1-weighted) images grant specificity for marrow abnormalities, sensitivity for muscle atrophy, characterization of blood products, and high-resolution depiction of structures that are surrounded by fat. Fluid-sensitive images (T2-weighted or STIR) are better used to identify the marrow and soft tissue edema that typify most pathology [1].

MR arthrography (MRA) is the current standard technique of advanced imaging for the evaluation of patients with suspected labral tears and chondral lesions. Limited data suggest that high-resolution unenhanced MRI, particularly performed at 3 T, may be as accurate as MR arthrography in the detection of labral and chondral tears. Large-FOV unenhanced MRI is far inferior to both MR arthrography and 3 T-unenhanced MRI in the detection of labral and chondral abnormalities of the hip [8]. A new technique called "traction MR arthrography" safely enables accurate detection and grading of labral and chondral lesions. Accuracy for detection of labral lesions reported was 92%/93%, 91%/83% for acetabular lesions, and 92%/88% for femoral cartilage lesions for reader 1 and reader 2, respectively [9].

4.2 Imaging of Common Hip and Groin Injuries in the Athlete

Sports-related injuries during the latest London Olympic Games had an overall injury rate of 12.9 per 100 athletes; types of injuries varied according to sport and athletes; soccer and ice hockey caused the greatest portion of injuries [10]. Cause of pain for discussion purposes can be summarized in **hip pathology** (e.g., femoroacetabular impingement, acetabular labral tears, cartilage pathology, loose bodies), **soft tissue injuries** including acute and chronic muscle-tendon pathology (anterior, medial, and lateral soft tissue injuries of the hip and groin), and **skeletal injuries** (acute/stress fractures, avulsions in skeletal immature athlete, osteitis pubis, and symphysis pubis arthropathy). Other differential diagnoses including **visceral pathology** (inguinal hernia, urogenital pathology) and **nerve entrapment syndrome pathology** should be considered in evaluating these athletes. This chapter examines the imaging features of the most common **acute** and **chronic** hip and groin injuries.

4.3 Hip Pathology

4.3.1 Femoroacetabular Impingement

In the athlete, FAI is a major cause of hip pain, reduced range of motion, and decreased performance. FAI has become increasingly recognized as a disorder that can lead to progressive labral and chondral injury and early hip degeneration.

The most common structural deformities that lead to dynamic mechanical overload are a loss of femoral head-neck offset ("cam" impingement, Fig. 4.2), a focal or global acetabular overcoverage ("pincer" impingement, Fig. 4.3), and a combination of impingement deformities [11].

The anatomic alterations underlying impingement may be secondary to another insult to the hip including developmental dysplasia, Legg-Calvé-Perthes disease, slipped capital femoral epiphysis, as well as posttraumatic and postoperative deformities [12–14].

Cam impingement is the consequence of an abnormal femoral head-neck relationship resulting in an osseous bump lateral or anterosuperior (1–2 o'clock) position using a clockface orientation on the femoral neck. It is more common than pincer impingement and is characteristically seen in young male athletes. Abnormal osseous morphology underlying cam impingement includes decreased femoral anteversion, abnormal femoral head-neck offset, shallow taper between the femoral head and neck, nonspherical femoral head, femoral neck "bump" or osseous excrescence, pistol grip deformity, and generalized enlargement of the femoral head (coxa magna) [15–18]. Conventional radiographs are very useful in identifying many of these anatomic alterations; measuring the alpha angle has been described to quantify the atypical head-neck correlation (Fig. 4.4) [19].

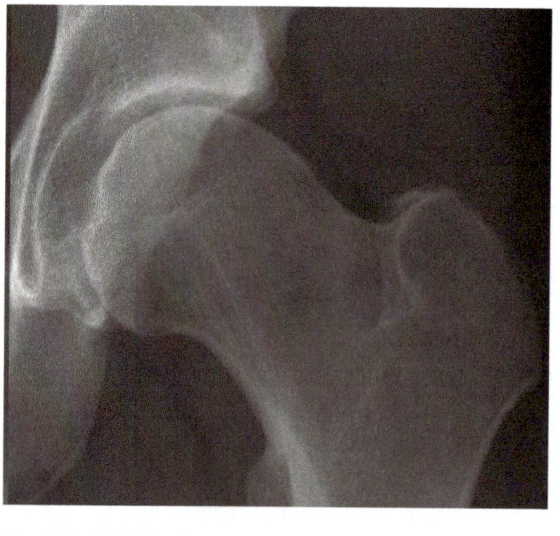

Fig. 4.2 RX AP view demonstrating cam FAI with nonspherical femoral head shape (pistol grip deformity)

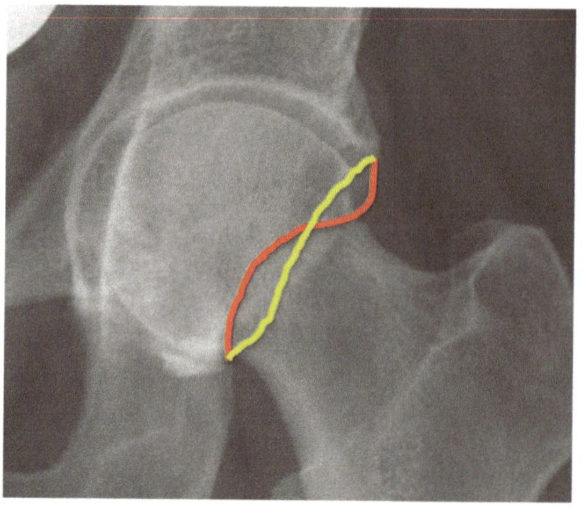

Fig. 4.3 RX AP view demonstrating pincer FAI with crossover sign (COS). The COS testifies the overlap between the anterior and the posterior wall of the acetabulum. The posterior acetabular wall is outlined in *yellow* and the anterior in *red*. In a hip with a normal anteversion, the line of the anterior wall remains medial to the line of the posterior wall; instead in acetabular retroversion, the line of the posterior wall crosses medially the line of the anterior wall

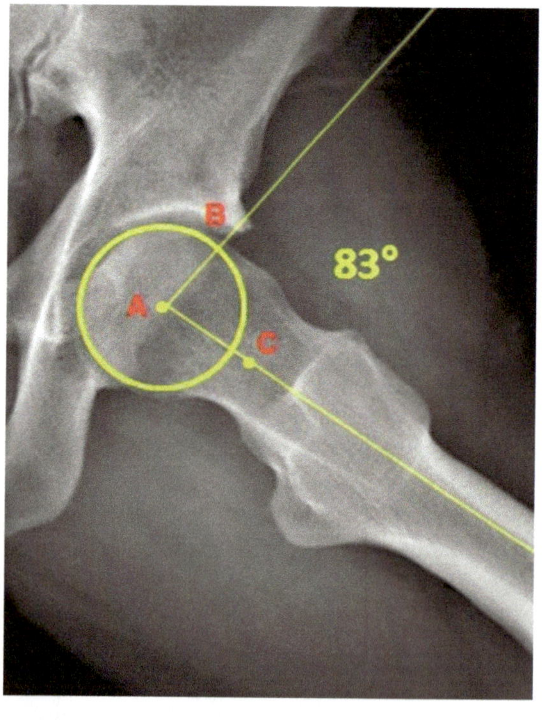

Fig. 4.4 The alpha angle. This angle is used for identifying an abnormal femoral head-neck junction. A best-fit circle is drawn, which outlines the femoral head. The angle is calculated as the angle formed between line that bisects the femoral neck and the point where the femoral head protrudes anterior to the *circle*. The angle ACB should be less than 50°

In this variety of impingement during movement of hip flexion, internal rotation, and adduction, bony outgrowth or bulge of the femoral neck avoids the normal movement of the femur within the acetabulum. This results in a characteristic pattern of shear injury to the transition zone and adjacent acetabular articular cartilage. Labral tears secondary to primary cam impingement typically result in detachment at the transition zone cartilage rather than intrasubstance injury. The cartilage break begins at the margin of the joint and progresses more centrally (outside to inside delamination) as the damage advances [20].

Pincer impingement is not as commonly seen as the cam mechanism; it occurs during flexion and internal rotation movements in middle-aged females. The underlying pathoanatomical condition is caused by acetabular overcoverage. Osseous impaction along the anterosuperior or superior femoral neck with compression proceeds with injury of the anterosuperior labrum and a small rim of chondromalacia. Posteroinferior cartilage lesions are caused by "contrecoup" forces in pincer-type impingement. This condition may be seen with excessive acetabular retroversion, protrusio acetabuli, and coxa profunda [21, 22]. The primary site of injury is the labrum. It is "pinched" between the femoral neck and the acetabular rim, and the damage is usually intrasubstance tearing of the labrum; the secondary site of injury is articular.

The diagnostic imaging evaluation protocol for FAI should include an AP pelvis radiograph and an elongated neck (Dunn) lateral view of the affected hip [23]. The AP pelvis has been demonstrated to be a valid indicator of cranial retroversion in the presence of a positive crossover sign; this is defined with the anterior rim line being lateral to the posterior rim in the cranial part of the acetabulum and crossing the latter in the distal part of the acetabulum [24]. The Dunn view (at either 45° or 90° of hip flexion) provides an improved evaluation of the femoral head-neck geometry. This view allows for identification of the cam morphology and calculation of an alpha angle, which estimates the degree of asphericity of the femoral head.

MRI with or without gadolinium contrast of the affected hip will allow accurate delineation of the intra-articular and periarticular soft tissue structures, including the femoral and acetabular chondral surface, labrum, capsule, and surrounding extra-articular tendinous insertions. Advanced cartilage imaging including T2 mapping, T1 rho, and delayed gadolinium-enhanced MRI of cartilage (dGEMRIC) techniques has allowed for improved quantitative assessment of cartilage integrity. Limited data suggest that high-resolution unenhanced MRI, particularly performed at 3 T, may be as accurate as MR arthrography in the detection of labral and chondral tears. Large-FOV unenhanced MRI is far inferior to both MR arthrography and 3 T-unenhanced MRI in the detection of labral and chondral abnormalities of the hip [8]. A new technique called "traction MR arthrography" safely enables accurate detection and grading of labral and chondral lesions. Accuracy for detection of labral lesions reported was 92%/93%, 91%/83% for acetabular lesions, and 92%/88% for femoral cartilage lesions for reader 1 and reader 2, respectively [9].

A CT scan with three-dimensional reconstruction is particularly useful for minimally invasive techniques such as arthroscopic or mini-anterior approaches to surgical management of impingement. It also helps to delineate complex combined mechanical pathomorphology. Post-processing computer analysis of three-dimensional CT imaging can allow for analysis of conflict patterns within the involved hip. Computer-assisted modeling of surgical corrections can be used for preoperative planning with improved accuracy of surgical intervention [7].

An ultrasound-guided intra-articular analgesic and steroid injection into the hip may be used for both diagnostic and therapeutic purposes and is an important adjunct to the overall evaluation. Response to an intra-articular injection has been shown to be 90% reliable as an indicator of an intra-articular abnormality.

4.3.2 Acetabular Labral Tear

The underlying cause of the labral injury should be delineated. There appear to be five common causes of labral tears: (1) trauma, (2) laxity/hypermobility, (3) bony impingement,

(4) dysplasia, and (5) degeneration [25]. The labrum is a rim of fibrocartilage that attaches to the base of the acetabular rim. It surrounds the perimeter of the acetabulum and is absent inferiorly where the transverse ligament resides. Labral tears can be traumatic or acute, chronic, and in some cases degenerative in nature. Loss of labral function can lead to overloading of the articular cartilage of the hip and may be a precursor of osteoarthritis in the hip. Associations between labral tearing and articular cartilage damage have been reported. The articular lesions are most often located adjacent to the labral tear, often at the labral-chondral junction [26]. Direct MRA has been shown to be the best imaging modality for assessment of labral pathology [27, 28]. The labrum is triangular in shape, with its base attached to the acetabular rim and its apex extending laterally along the capsular side of the acetabular rim. The labrum demonstrates typical MRI features of organized collagen, with decreased low signal on T1- and T2-weighted images [29].

Most labral pathology occurs in the anterior and anterosuperior aspect of the joint. Tears are divided into three major groups: detached, intrasubstance, and degenerative. Labral detachments involve separation of the labral from the acetabular rim and are identified by contrast material interposed between the labrum and the rim (Fig. 4.5).

Detachments may be complete or partial, with or without a displaced fragment, and are best seen on coronal images in the setting of superior predominant tears and oblique axial images for anterior predominant tears.

Intrasubstance labral tears classically demonstrate intrasubstance fluid or contrast signal, usually extending to the articular side of the labrum, which is often oblique or curvilinear in shape. However, the signal may also be complex and extend in multiple directions in the long and short axis of the labrum. Increased signal alone within the labrum can be seen without frank fluid or contrast signal in the setting of labral tears with opposed surfaces [30].

A labrum with abnormal irregular contours and a thin morphology, with or without intrasubstance fluid or a contrast signal that extends to the free margin, is considered a degenerative-type tear. In athletes, it is not uncommon to have

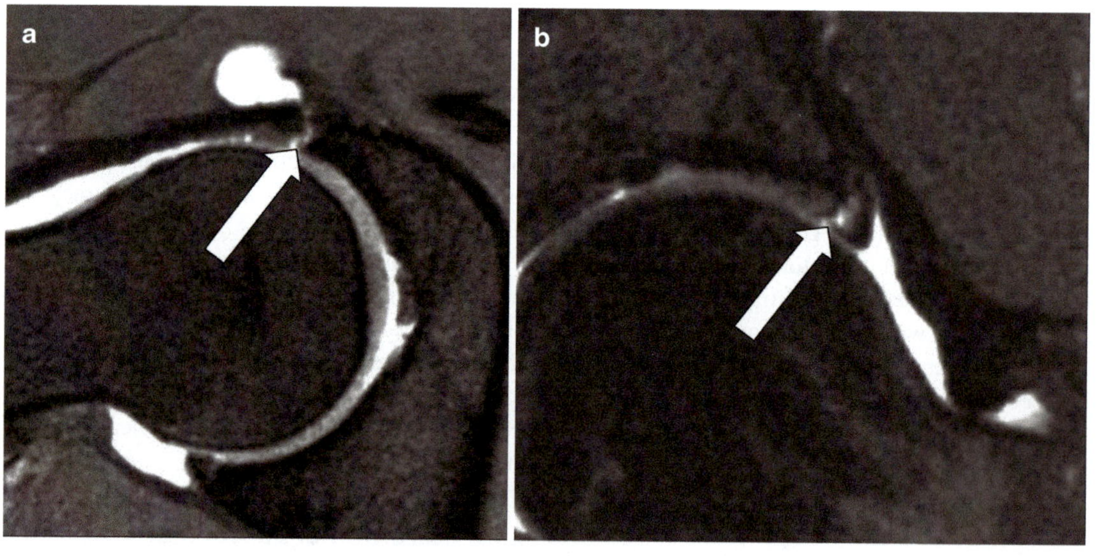

Fig. 4.5 MRI arthrogram of the hip demonstrating labral tears. (**a**) The *white arrow* demonstrates where the labrum is torn away from the acetabulum with well-defined fluid intensity lesion formation (anterior paralabral cyst). (**b**) The *white arrow* demonstrates where the labrum is partially torn away from the acetabulum

a combination of detached or intrasubstance tears with superimposed degenerative components.

4.3.3 Loose Bodies

Loose bodies within the joint can cause pain and may mimic the snapping hip phenomenon. Anterior groin pain, episodes of clicking or locking, buckling, giving way, and persistent pain during activity suggest an intra-articular loose body. Loose bodies within the hip, whether ossified or not, have been correlated with locking episodes and inguinal pain [31]. Besides hip trauma, other diseases known to be associated with loose bodies include Legg-Calvé-Perthes disease, osteochondritis dissecans, avascular necrosis, synovial chondromatosis, and osteoarthritis. Imaging with CT scan versus MRI should be thoughtfully assessed depending on the most likely diagnosis. If the loose particles are more likely to be chondral or soft tissue, then MRI is more likely to yield useful diagnostic information. The use of cartilage-sensitive MRI sequences (e.g., T2 mapping, T1 rho, and delayed gadolinium-enhanced MRI of cartilage (dGEMRIC) techniques) for the detection of lesions, therefore, may be preferable, although CT scans are more sensitive than MRI for detection of suspected bony loose bodies [32].

4.4 Soft Tissue Injuries of the Hip and Groin in the Athlete

4.4.1 Adductor-Rectus Femoris—Hamstring Injuries

Acute muscle injury occurs as a result of two mechanisms. Direct blunt trauma or compressive injury results in contusions and hematomas. Movements, where the muscle is being stretched through forceful contraction, lead to muscle strains and ruptures.

In most cases, the lesion is close to the musculotendinous junction, but in some cases, the tendon itself or the enthesis where the tendon inserts into the bone is the site of the injury. The three most common acute musculotendinous injuries in the hip and groin region occur in the adductor muscles (usually the adductor longus), the hamstring muscles (most often biceps femoris), and the quadriceps muscle (most often the rectus femoris).

In other cases, hip and groin injuries have the characteristics of an overuse injury, and the exact moment of injury can be hard to establish.

Adductor strains (pectineus, brevis, longus, magnus, and gracilis) occur primarily in the proximal portion of the adductor longus, at the muscle-tendon junction, or at the insertion of the pubic bone itself. The adductor longus and gracilis muscle enthesis are the most affected (Fig. 4.6a, b).

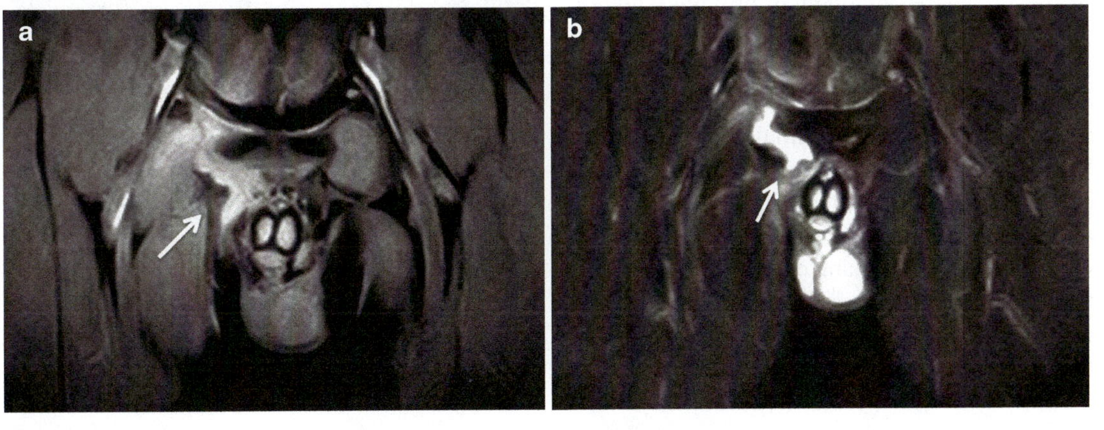

Fig. 4.6 (**a**) MRI coronal FWGE SAT and (**b**) MRI coronal STIR demonstrating complete detachment of adductor longus muscle enthesis with hematoma formation

The microtearing pattern is more often related to the development of chronic groin pain but may come early as complete rupture. In low-grade injuries, US has demonstrated a low sensitivity for detecting partial strain or tenoperiosteal disease. In these cases, MR imaging displays more reliably the pathoanatomy process. On the other hand, the US diagnosis of complete detachment of both muscles from their insertions is simple.

The rectus femoris has two heads, the direct originating from the anteroinferior iliac spine (AIIS) and the indirect originating from the superior acetabular rim and hip capsule. Proximal tears occur less frequently than tears in the midsubstance of the muscle belly involving the central aponeurosis or at the distal myotendinous junction. Rectus femoris tendon ruptures either at its insertion into the bone with detachment of bone fragment (avulsion from the apophysis in skeletally immature patients) or at the proximal myotendinous junction involving its direct head. In adolescent, acute pain proximally at the muscle origin indicates an apophyseal injury and must be high on the checklist of differential diagnosis. Partial tendon tears are more common than complete detachments. Tears affecting the distal insertion of the muscle fibers into the deep distal aponeurosis seem more common (Fig. 4.7a, b) [33].

The proximal hamstring complex has a strong bony attachment on the ischial tuberosity. The footprint on the ischium is composed of the semitendinosus and the long head of the biceps femoris beginning as a common proximal tendon footprint, with a distinctly separate semimembranosus footprint. Hamstring tendons can be injured following chronic microtrauma or a single acute injury. The proximal insertion of the long head of the biceps femoris and the semitendinosus is more commonly involved than that of the semimembranosus (Fig. 4.8a, b).

There is a continuum of hamstring injuries that can range from musculotendinous strains to avulsion injuries. By definition, a strain is a partial or complete disruption of the musculotendinous unit. A complete tear or avulsion, in contrast, is a discontinuity of the tendon-bone unit [34].

There is a recognized grading system for muscle injuries established by Munich consensus statement [35]. Injuries have been classified as structural or nonstructural according to presence of anatomically evident lesion or not, respectively.

Plain films may be helpful to rule out fractures or bony avulsion injuries but are often negative. Ultrasound (US) is usually the first-level diagnostic examination, which allows staging of almost all muscle lesions and assessment of their evolution and complications. The most sensitive and the gold standard for imaging muscle strains is magnetic resonance imaging (MRI); however, the most recent evidences recommend to combine MRI and US. On MRI, muscle strains are graded as first, second, or third degree, depending on the involvement of the muscle fibers. First-degree strains are characterized by microscopic injury, <5% fiber disruption, and minimal edema/hemorrhage track-

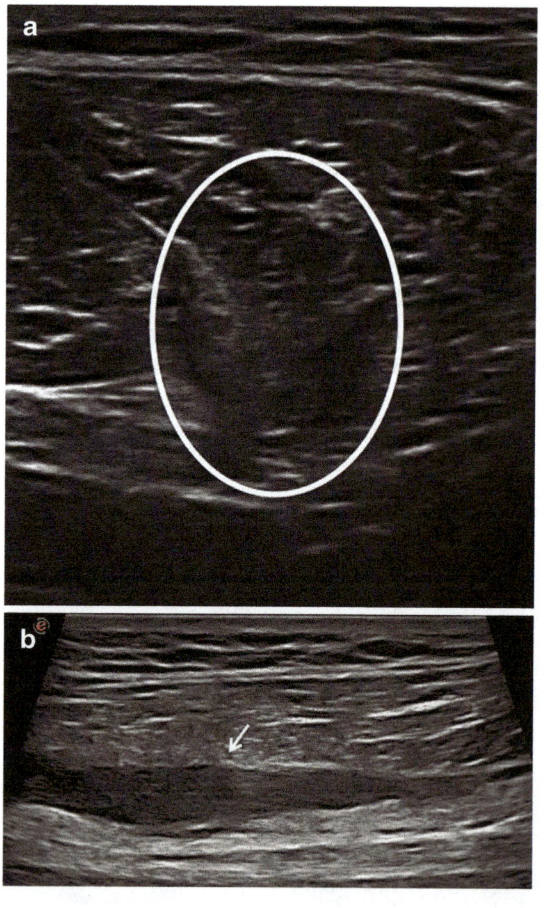

Fig. 4.7 (**a**) US short-axis rectus femoris and (**b**) US long-axis rectus femoris. Central aponeurosis tears of the rectus femoris with large hypoechoic area (*circle*) associated with anechoic fluid collection (*arrow*) reflecting hematoma formation

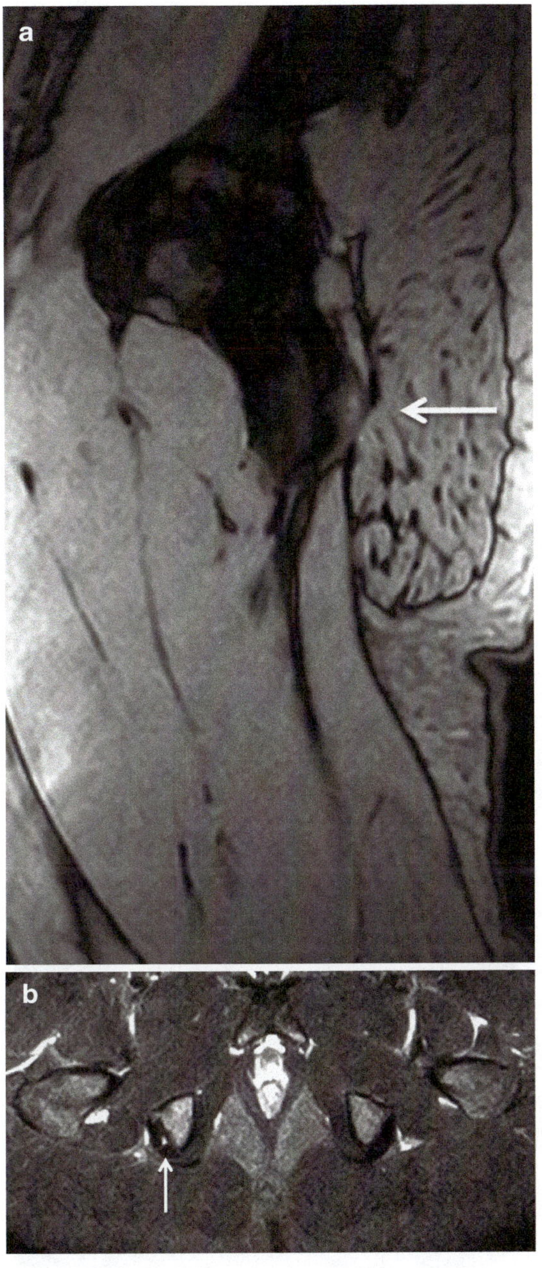

Fig. 4.8 (**a**) MRI sagittal FWGE T2* and (**b**) MRI axial STIR slight thickening (*arrow*) of proximal insertion of the long head of the biceps femoris and the semitendinosus with small partial hyperintense intrasubstance tear (*arrow*)

ing along muscle fascicles. Second-degree strains involve partial thickness and show up as high-signal intensity on T2-weighted images within the muscle with perifascial fluid present. Third-degree strains show complete musculotendinous disruption with or without retraction [36].

4.4.2 Quadriceps Contusions: Hip Pointers—Morel-Lavallée Lesion

Contusions are caused by direct blunt trauma; the muscle belly is compressed against the underlying bone; some common sites of injury include iliac crest, proximal thigh, and oblique muscles of the abdominal wall. These injuries are often clinically diagnosed without difficulty, but US and MRI are often used when patients symptoms are not resolving as expected. On imaging, there is no typical myotendinous junction localization as seen in muscle strain. US is regarded less sensitive compared to MRI [37].

MR images reveal edema at the injured site, frequently due to interstitial hemorrhage as well as edema. Typically, there are also skin edema and sometimes bone contusion. The signal intensity of a hematoma on T1W and T2W images depends on the degradation of hemoglobin. The major complications of muscle contusion are myositis ossificans and seroma [38].

Hip pointers are a result of a contusion to the iliac crest, which is protected only by a layer of subcutaneous fat. Typically these athletes will have significant pain radiating to the associated abdominal obliques proximally and/or abductors distally. The injury is typically self-limiting.

The Morel-Lavallée lesion is an injury that results in degloving of the skin and subcutaneous tissue from the neighboring fascia. Around the hip and pelvis, this most frequently involves the peritrochanteric region and results in a blood-filled cavity (Fig. 4.9a, b). US can nicely evaluate this lesion and show the extension because of its superficial localization; US-guided aspiration of fluid followed by local compression helps to prevent local recurrence.

4.4.3 Weakness of the Inguinal Canal Wall

The symphysis pubis represents the crossroads for many of the structures acting on the anterior pelvis. It is the common attachment of the confluence of the rectus abdominis fascial sheath with part of the adductor longus tendon fibers and with part of a common, fused tendinous origin of the gracilis and adductor brevis [39]. The rectus abdominis and the above tendon fibers described form a conjoined aponeurosis (common aponeurotic plate or "prepubic aponeurotic complex") which adheres to the central portion of the anterior pubic surfaces and also to the inguinal ligament at its insertion into the pubic tubercle and its prepubic extension. The fibers of the aponeurosis blend with both the fibrocartilaginous disc and the pubic bone hyaline cartilage anteriorly [40]. The inguinal ligament does not terminate at the pubic tubercle but extends obliquely over the anterior surface of the symphysis pubis. In only 3–5% of groins, a conjoint tendon represents the fused inferomedial aponeuroses of the internal oblique and transversus abdominis and inserts into the pubic tubercle and adjacent pubic crest. The inferior edge of the external oblique aponeurosis forms the inguinal ligament. The internal oblique and transversalis fascia comprise the posterior abdominal wall and join with the deep rectus sheath; a defect in the transversalis fascia corresponds to the internal inguinal ring [41].

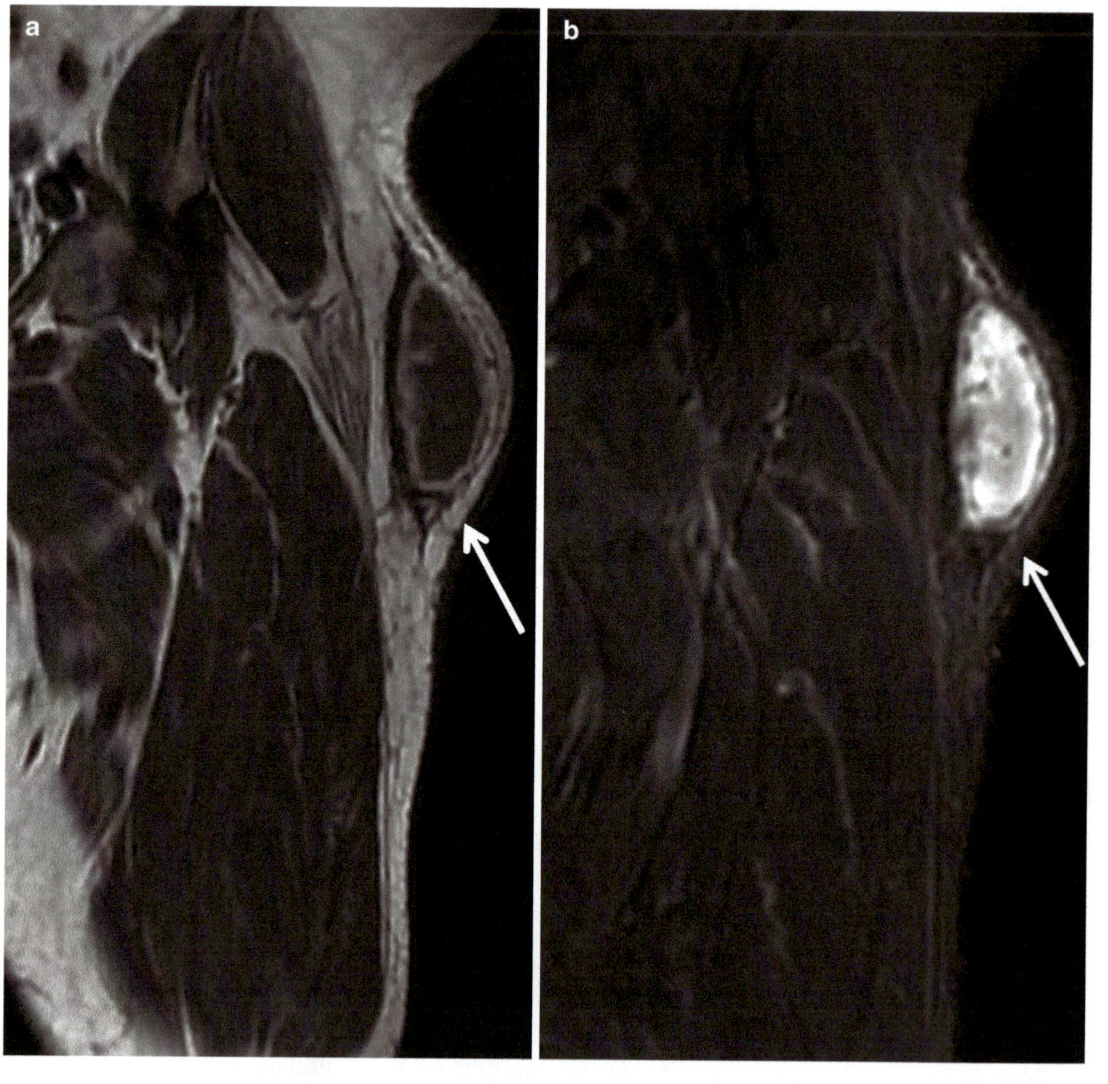

Fig. 4.9 (**a**) MRI sagittal T1 and (**b**) MRI sagittal STIR. This lesion is characterized by a filled cystic cavity (*arrow*) created by separation of the subcutaneous tissue from the underlying fascia in classic location over the greater trochanter

Therefore, an injury involving the common aponeurotic plate, the external oblique aponeurosis, and the internal oblique/transversalis fascia (posterior abdominal wall) alters the biomechanics disrupting the anatomic continuity of these connective tissues, and, in turn, such disruption leads to instability of the symphysis pubis.

The concept of the anatomic and functional relationship between the rectus abdominis, the adductors, the inguinal canal, and the symphysis pubis provides a unifying concept for groin injuries in athletes. The result is a vicious circle of joint micro-macro instability (e.g., sacroiliac, femoroacetabular, and symphysis pubis) and tendon-muscular imbalance. Possible sources of pain have been theorized to be the result of entrapment of the genital branches of the ilioinguinal or genitofemoral nerves. The symphysis itself is innervated by branches of the pudendal and genitofemoral nerves. Other reports have suggested that the iliohypogastric or obturator nerves could potentially be involved [42, 43].

Either due to shearing injury from below or traction injury from above, the final effect will be a lesion of the common aponeurotic plate with its detachment from the underlying pubic bone. MRI is the modality of choice in the depiction of common adductor-rectus abdominis and aponeurotic injuries; the radiologic findings of major interest can be summarized as follows [44]:

- The presence of grade 2–3 bone marrow edema (BME) at the symphysis pubis level (Fig. 4.10a, b)

The presence of BME must be identified into the sequences coronal STIR, coronal T1, axial oblique T2 FS, and PD FS. Furthermore, BME should also be classified into I, II, or III degree, in relation to its extension measured into the PD FS or T2 FS axial oblique plan sequences (grade 0, no BME; grade 1, BME ≤1 cm; grade 2, BME ≥1 cm and ≤2 cm; grade 3, BME ≥2 cm). A grade 1 BME could be normally present due to functional adaptation specific to high-level performance model of these athletes.

- The presence of secondary inferior and/or superior cleft sign must be assessed in coronal STIR, axial oblique PD FS, and sagittal STIR sequences. This useful imaging sign is a curved, high-signal intensity line along the medial corner of the pubic bone on a coronal water-sensitive image (Fig. 4.11a–c), a finding that is not present in control subjects and represents aponeurotic complex injury (Fig. 4.12a–c) with tear or partial avulsion of adductor longus origin [45].
- Evidence of symphysis central disc protrusion: this sign should be estimated in coronal T1 and axial oblique T1 sequences (Fig. 4.13).
- Adductor longus muscle-tendon injury: this lesion should be evaluated in axial oblique sequences PD FS and T2 FS, as well as coronal STIR images (Fig. 4.14a, b).

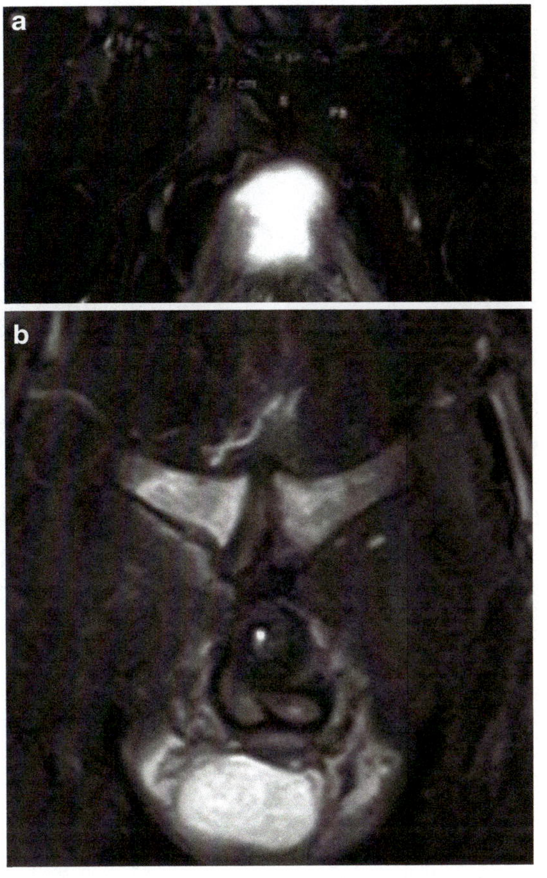

Fig. 4.10 (**a**) MRI axial STIR and (**b**) MRI coronal STIR. The *first image* shows right monolateral grade 3 BME, extending from the anterior to the posterior edge of pubic bone. The *second image* demonstrates bilateral grade 3 BME

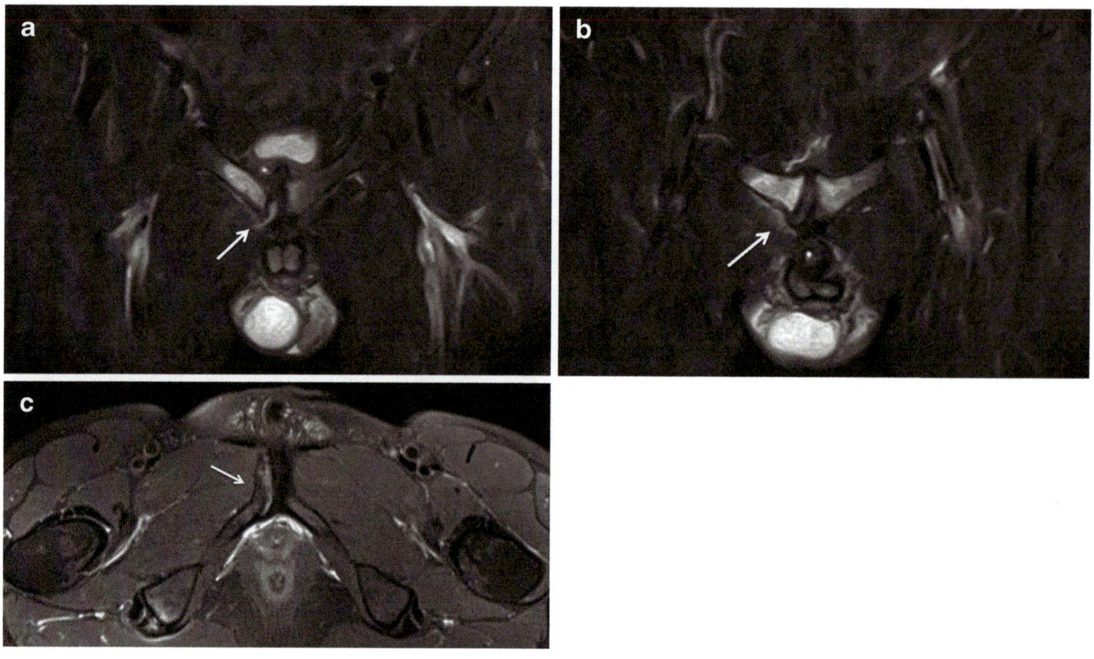

Fig. 4.11 (**a**) MRI coronal STIR, (**b**) MRI coronal STIR, and (**c**) MRI axial STIR. These images show hyperintense line along the inferior border of the pubic bone, communicating with the articular surface

Imaging findings of posterior inguinal wall deficiency have been described on dynamic sonography. A combined technique involving the radiologist and the surgeon during the ultrasound examination is preferred by the author. While the examining surgeon's finger follows the spermatic cord upward above the inguinal ligament the external inguinal ring, the athlete is asked to strain with the probe placed over the medial aspect of the inguinal region. The protocol provides scanning the inguinal canal both in long- and short-axis views (Fig. 4.15).

The diagnosis of a "true" inguinal hernia is confirmed if an "impulse" or bulge is felt by the examination finger and detected by the scan; the examination is positive for posterior inguinal wall deficiency if abnormal ballooning of the posterior inguinal wall is visible during the scan and not felt by the examination finger.

4.5 Skeletal Injuries

4.5.1 Apophyseal Avulsions of the Hip and Pelvis in Adolescents

The apophyseal growth plate is weaker than tendons, ligaments, and muscles in these growing athletes and is therefore the site of failure or injury. The locations for hip and pelvic apophysis and respective muscle groups include the ischial tuberosity (proximal hamstrings), anteroinferior iliac spine (rectus femoris), anterosuperior iliac spine (sartorius), lesser trochanter (psoas), iliac crest (abdominal obliques), inferior pubic ramus (adductors), pubic tubercle (rectus abdominis), and greater trochanter (gluteus medius and minimus). The most common sites for apophyseal avulsions are the ischial tuberosity (54%), anteroinferior iliac spine (22%), and anterosuperior iliac spine (19%), followed by the pubic

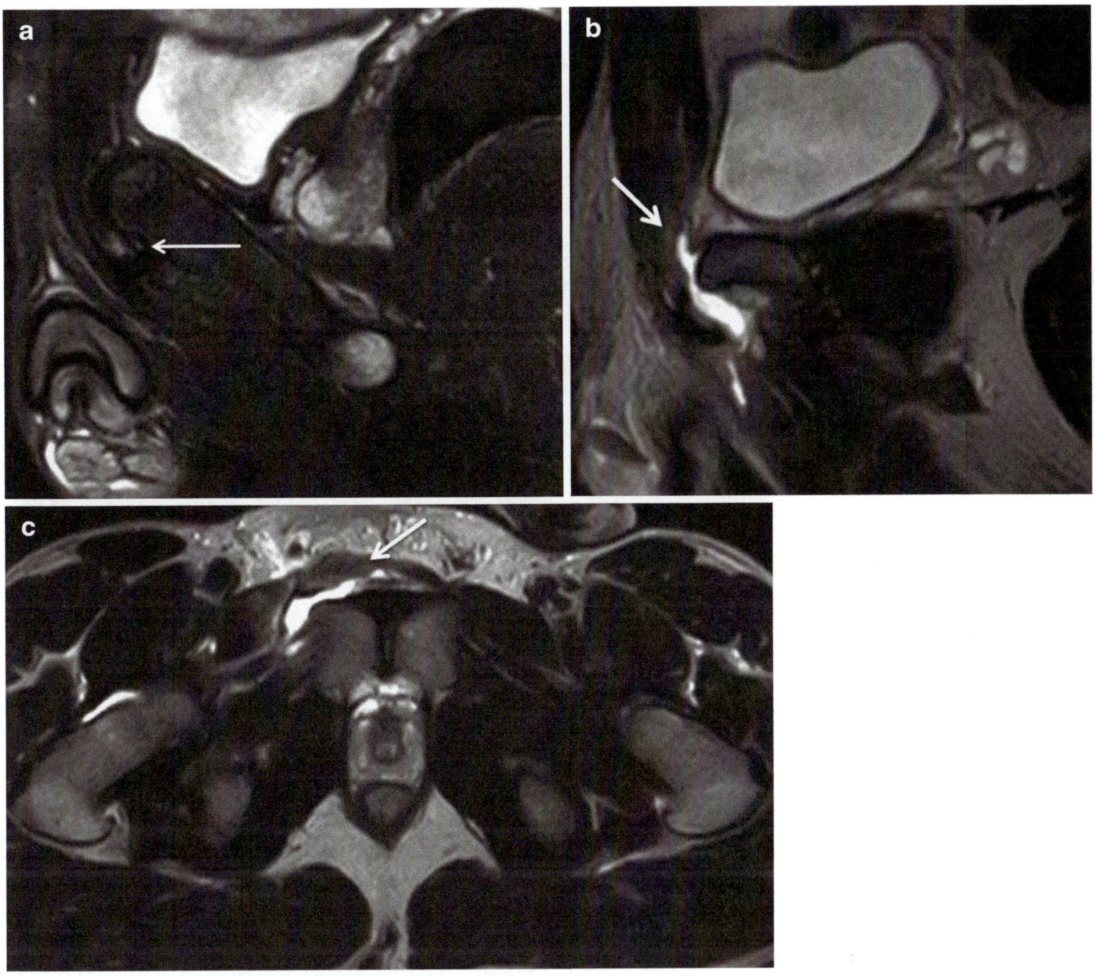

Fig. 4.12 (**a**) MRI sagittal STIR, (**b**) MRI sagittal STIR, and (**c**) MRI axial T2. The *first image* shows a small hyperintense focal area under the pre-aponeurotic complex, representing a small injury. The *second* and *third images* show a bigger area of detachment of the pre-aponeurotic complex from the pubic bone

tubercle and iliac crest [46]. These injuries are typically recognized on radiographs with separation of the apophysis from the pelvis. In some cases, however, the avulsion does not result in significant apophyseal displacement, and the injury is not easily recognized radiographically. Magnetic resonance imaging is useful for these patients in evaluating signs of growth plate widening with hyperintense signal on fluid-sensitive images. Edema may be present in the adjacent bone, periosteum, and tendon origin (Fig. 4.16a, b) [47].

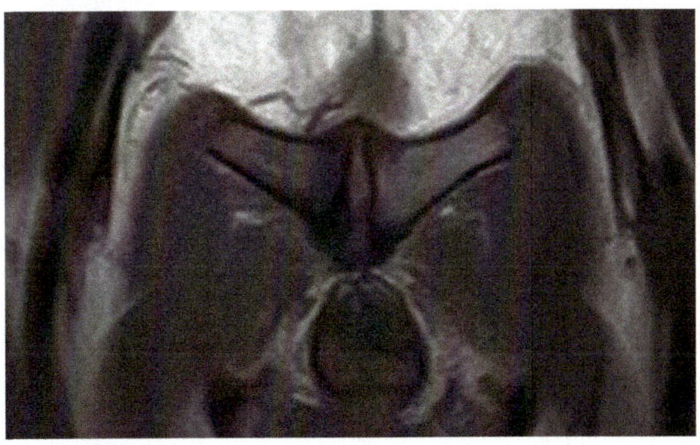

Fig. 4.13 MRI coronal T1. The image shows a bulge of the symphysis central disc with signs of pubic arthropathy

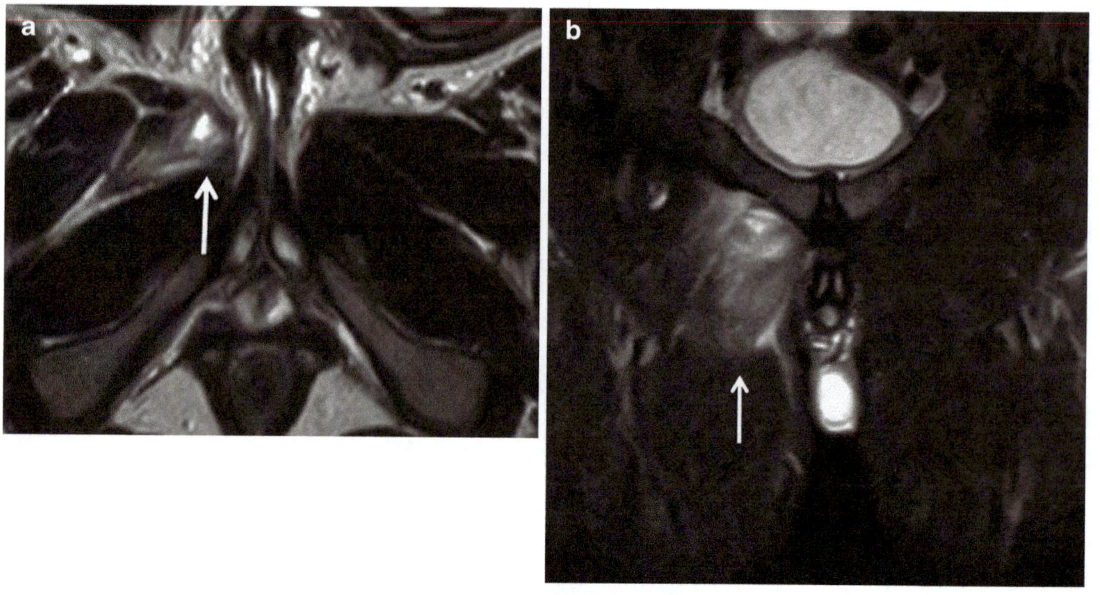

Fig. 4.14 (a) MRI axial T2 and (b) MRI coronal STIR. These images demonstrate a grade 2 injury to adductor longus muscle-tendon unit, with continuity disruption of the proximal third muscle belly fibers and hematoma formation

4.5.2 Stress Fractures of the Hip and Pelvis

The term "bone stress injury" describes a spectrum of injuries to the bone that encompasses both **stress reactions** and **stress fractures**. A combination of increased mechanical strain and increased rate of bone loading may stress the bone beyond its normal repair capacity. Without adequate time to heal, this leads to damage accumulation and fatigue failure of the bone. Stress fractures are often described as either fatigue fractures or insufficiency fractures. Fatigue fractures occur from repetitive strain over a prolonged time to normal bone structure. Insufficiency fractures occur as a result of normal stress load to an abnormal bone structure [48, 49]. They are generally described as tension-sided (lateral femoral neck) or compression-sided (medial femoral neck) fractures. Typical locations are the femur shaft (53%) (Fig. 4.17a, b), lesser trochanter (20%), intertrochanteric (15%), femoral neck (11%), and trochanter major (1%) [50]. Radiographs are often the first imaging modality, but, unfortunately, most of the changes visible on plain film radiographs (periosteal elevation, endosteal elevation, cortical sclerosis, and subtle blurring of the trabecular margins) are not apparent in the early stages. Magnetic

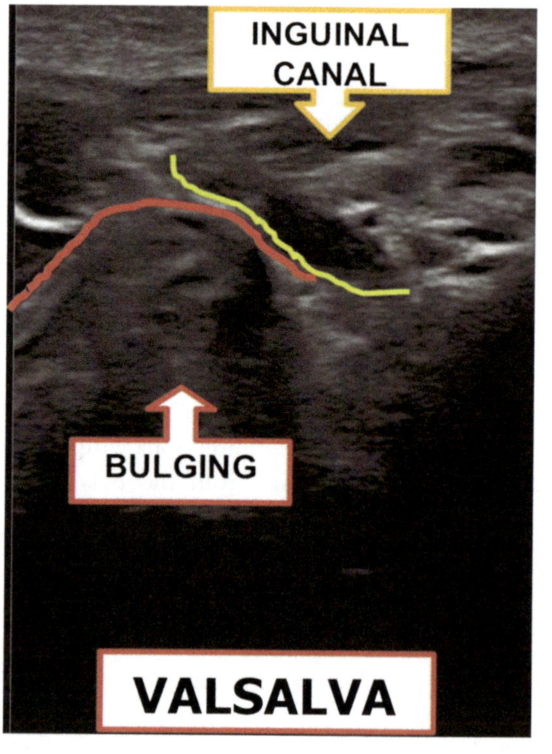

Fig. 4.15 US dynamic examination. During "tense maneuver" of the abdominal muscles, it is possible to identify an incipient bulging (*red line*) of the posterior inguinal wall, conflicting with structures of the inguinal canal

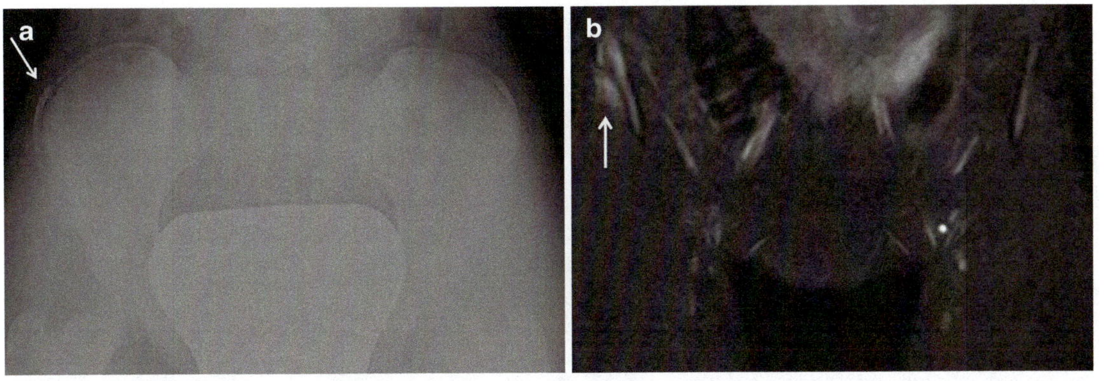

Fig. 4.16 (a) RX AP pelvis and (b) MRI coronal STIR. These images demonstrate a slight displacement of the right superior iliac crest, with respect to the contralateral side. Edema of the anterior abdominal wall (muscle belly) attaching on the iliac crest is visible as diffuse area of signal hyperintensity on this fluid-sensitive sequence

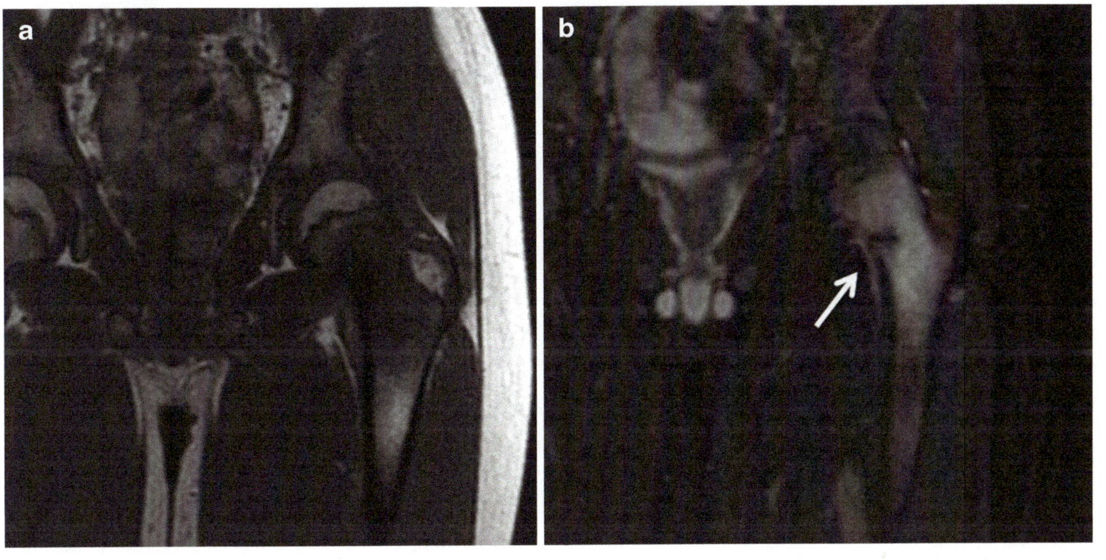

Fig. 4.17 (a) MRI coronal T1 and (b) MRI coronal STIR. A true fracture line can be seen as a line of low signal intensity continuous from the cortex to the intramedullary space surrounded by an area of bony edema visible on fluid-sensitive sequence

resonance imaging (MRI) is currently the imaging modality of choice for evaluation of most stress injuries, it also allows detailed evaluation of surrounding soft tissues, which may be helpful in evaluating other or concomitant injuries.

4.5.3 Osteitis Pubis

Osteitis pubis is characterized by pain, instability, and bony changes in the symphysis pubis [51]. The symphysis pubis often demonstrates radiographic and MRI abnormalities in both symptomatic and asymptomatic athletes. The term *osteitis pubis* is frequently used when referring to the radiographically abnormal symphysis pubis, although originally was used to describe an infection at the pubic bone at the symphysis joint level, and it should be more appropriately termed pubic bone stress injury, as is generally due to overuse and overload in athletes where the muscular control around the pelvis is poorly coordinated or imbalanced.

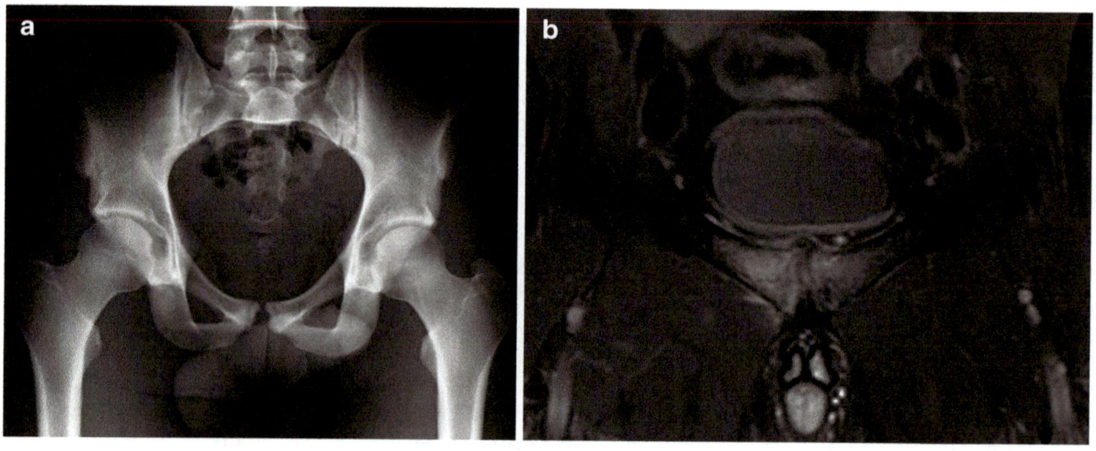

Fig. 4.18 (**a**) RX AP pelvis and (**b**) MRI coronal STIR. Signs of arthrosis are visible in this RX study, associated with bilateral BME on MRI imaging

Chronic instability of the symphysis pubis resulting from repetitive microtrauma shows typical signs of arthrosis such as joint space narrowing or movement greater than 2 mm which is evident on flamingo views (views with weight bearing on a single leg), subchondral sclerosis and cysts, as well as productive changes such as bony spurs and enthesopathy of the adductors, whereas active chronic pubic osteitis exhibits appositional bone marrow edema (Fig. 4.18a, b) [52].

Conclusions

A full knowledge of the osseous and soft tissue anatomy, kinematics, and pathology of the hip and groin is essential to perform a correct imaging workup of athletes with sports injuries because there are numerous differential diagnostic causes of groin pain; both static and dynamic imaging modalities may contribute to reveal primary and secondary injuries.

Targeted RX, MRI, and ultrasound examination protocols play a key role in accurate diagnosis and therapy planning for these athletes. Close interrelationship between the radiologist and the team of sports medicine specialists is needed to concentrate the imaging examinations within a fine clinical context.

References

1. Omar IM, Zoga AC, Kavanagh EC, Koulouris G, Bergin D, Gopez AG, et al. Athletic pubalgia and "sports hernia": optimal MR imaging technique and findings. Radiographics. 2008;28(5):1415–38.
2. Thorborg K, Serner A, Petersen J, Madsen TM, Magnusson P, Holmich P. Hip adduction and abduction strength profiles in elite soccer players: implications for clinical evaluation of hip adductor muscle recovery after injury. Am J Sports Med. 2011;39(1):121–6.
3. Death AB, Kirby L, Mc ML. Pelvic ring mobility: assessment by stress radiograph. Arch Phys Med Rehabil. 1982;63:204–6.
4. Agricola R, Waarsing JH, Thomas GE, Carr AJ, Reijman M, Bierma-Zeinstra SM, Glyn-Jones S, Weinans H, Arden NK. Response to letter to the editor: "Cam impingement: defining the presence of a cam deformity by the alpha angle data from the CHECK cohort and Chingford cohort". Osteoarthritis Cartilage. 2014;22(12):2095–6.
5. Peetrons P. Ultrasound of muscles. Eur Radiol. 2002;12:35–43.
6. Dawes AR, Seidenberg PH. Sonography of sports injuries of the hip. Sports Health. 2014;6(6):531–8.
7. Bedi A, Dolan M, Magennis E, Lipman J, Buly R, Kelly BT. Computer-assisted modeling of osseous impingement and resection in femoroacetabular impingement. Arthroscopy. 2012;28(2):204–10.
8. Naraghi A, White LM. MRI of labral and chondral lesions of the hip. Am J Roentgenol. 2015;205:479–90.
9. Schmaranzer F, Klauser A, Kogler M, Henninger B, Forstner T, Reichkendler M, Schmaranzer E.

Diagnostic performance of direct traction MR arthrography of the hip: detection of chondral and labral lesions with arthroscopic comparison. Eur Radiol. 2015;25(6):1721–30.
10. Engebretsen L, Soligard T, Steffen K, et al. Sports injuries and illnesses during the London Summer Olympic Games 2012. Br J Sports Med. 2013;47:407–14.
11. Philippon MJ, Stubbs AJ, Schenker ML, et al. Arthroscopic management of femoroacetabular impingement: osteoplasty technique and literature review. Am J Sports Med. 2007;35:1571–80.
12. Noguchi Y, Miura H, Takasugi S, et al. Cartilage and labrum degeneration in the dysplastic hip generally originates in the anterosuperior weight-bearing area: an arthroscopic observation. Arthroscopy. 1999;15: 496–506.
13. Eijer H, Myers SR, Ganz R. Anterior femoroacetabular impingement after femoral neck fractures. J Orthop Trauma. 2001;15:475–81.
14. Leunig M, Podeszwa D, Beck M, et al. Magnetic resonance arthrography of labral disorders in hips with dysplasia and impingement. Clin Orthop. 2004;418: 74–80.
15. Ito K, Minka II MA, Leunig M, et al. Femoroacetabular impingement and the cam effect: an MRI-based quantitative anatomical study of the femoral head-neck offset. J Bone Joint Surg Br. 2001;83:171–6.
16. Jager M, Wild A, Westhoff B, et al. Femoroacetabular impingement caused by a femoral osseous head-neck bump deformity: clinical, radiological, and experimental results. J Orthop Sci. 2004;9:256–63.
17. Stulberg SD, Cordell LD, Harris WH, et al. Unrecognized child- hood hip disease: a major cause of idiopathic osteoarthritis of the hip. In: The hip. Proceedings of the third meeting of the hip society. St. Louis: CV Mosby; 1975. p. 212–28.
18. Tonnis D, Heinecke A. Acetabular and femoral anteversion: relationship with osteoarthritis of the hip. J Bone Joint Surg Am. 1999;81:1747–70.
19. Notzli HP, Wyss RG, Stoecklin CH, et al. The contour of the femoral head-neck junction as a predictor for the risk of anterior impingement. J Bone Joint Surg Br. 2002;84:556–60.
20. Beck M, Kalhor M, Leunig M, et al. Hip morphology influences the pattern of damage to the acetabular cartilage: femoroacetabular impingement as a cause of early osteoarthritis of the hip. J Bone Joint Surg Br. 2005;87:1012–8.
21. Ganz R, Parvizi J, Beck M, et al. Femoroacetabular impingement a cause for osteoarthritis of the hip. Clin Orthop Relat Res. 2003;417:112–20.
22. Siebenrock KA, Schoeniger R, Ganz R. Anterior femoro-acetabular impingement due to acetabular retroversion. J Bone Joint Surg Br. 2003;85:278–86.
23. Meyer DC, Beck M, Ellis T, Ganz R, Leunig M. Comparison of six radiographic projections to assess femoral head/neck asphericity. Clin Orthop Relat Res. 2006;445:181–5.
24. Jamali AA, Mladenov K, Meyer DC, et al. Anteroposterior pelvic radiographs to assess acetabular retroversion: high validity of the "cross-over-sign". J Orthop Res. 2007;25(6):758–65.
25. Kelly BT, Buly RL. Hip arthroscopy update. HSS J. 2005;1:40–8.
26. McCarthy J, Noble P, Aluisio FV, Schuck M, Wright J, Lee JA. Anatomy, pathologic features, and treatment of acetabular labral tears. Clin Orthop. 2003;406:38–47.
27. McSweeney SE, et al. Hip and groin pain in the professional athlete. Can Assoc Radiol J. 2012;63:87–99.
28. Philippon MJ, Kuppersmith DA, Wolff AB, et al. Arthroscopic findings following traumatic hip dislocation in 14 professional athletes. Arthroscopy. 2009;25:169–74.
29. Armfield DR, Towers JD, Robertson DD. Radiographic and MR imaging of the athletic hip. Clin Sports Med. 2006;25:211–39.
30. Overdeck KH, Palmer WE. Imaging of hip and groin injuries in athletes. Semin Musculoskelet Radiol. 2004;8:41–53.
31. Petersen W, Petersen F, Tillman B. Structure and vascularization of the acetabular labrum with regard to the pathogenesis and healing of labral lesions. Arch Orthop Trauma Surg. 2003;123(6):282–8.
32. Connell DA, Potter HG, Wickiewicz TL. Noncontrast magnetic resonance imaging of superior labral lesions. 102 cases confirmed at arthroscopic surgery. Am J Sports Med. 1999;27:208–13.
33. Armfield DR, Kim DH, Towers JD, Bradley JP, Robertson DD. Sports-related muscle injury in the lower extremity. Clin Sports Med. 2006;25(4): 803–42.
34. Slavotinek JP, Verrall GM, Fon GT. Hamstrings injury in athletes: using MR imaging measurements to compare extent of muscle injury with amount of time lost from competition. AJR Am J Roentgenol. 2002;179:1621–8.
35. Mueller-Wohlfahrt HW, Haensel L, Mithoefer K, Ekstrand J, English B, McNally S, Orchard J, van Dijk CN, Kerkhoffs GM, Schamasch P, Blottner D, Swaerd L, Goedhart E, Ueblacker P. Terminology and classification of muscle injuries in sport: the Munich consensus statement. Br J Sports Med. 2013;47(6): 342–50.
36. Boutin RD, Fritz RC, Steinbach LS. Imaging of sports-related muscle injuries. Radiol Clin North Am. 2002;40(2):333–6.
37. Draghi F, Zacchino M, Canepari M, Nucci P, Alessandrino F. Muscle injuries: ultrasound evaluation in the acute phase. J Ultrasound. 2013;16(4):209–14.
38. Aroven JG, Garrick JG, Chronister RD, McDevitt ER. Quadriceps contusions: clinical results of

immediate immobilization in 120 degrees of knee flexion. Clin J Sport Med. 2006;16(5):383–7.
39. Davis JA, Stringer MD, Woodley SJ. New insights into the proximal tendons of adductor longus, adductor brevis and gracilis. Br J Sports Med. 2012; 46(12):871–6.
40. Robinson P, Salehi F, Grainger A, Clemence M, Schilders E, O'Connor P, et al. Cadaveric and MRI study of the musculotendinous contributions to the capsule of the symphysis pubis. AJR Am J Roentgenol. 2007;188(5):W440–5.
41. Condon RE. Surgical anatomy of the transversus abdominis and transversalis fascia. Ann Surg. 1971;173:1.
42. Akita K, Niga S, Yamato Y, Muneta T, Sato T. Anatomic basis of chronic groin pain with special reference to sports hernia. Surg Radiol Anat. 1999;21(1):1–5.
43. Davies AG, Clarke AW, Gilmore J, et al. Review: imaging of groin pain in the athlete. Skeletal Radiol. 2010;39:629–44.
44. Branci S, Thorborg K, Bech BH, et al. MRI findings in soccer players with long-standing adductor-related groin pain and asymptomatic controls. Br J Sports Med. 2015;49:681–91.
45. Zoga AC, Kavanagh EC, Omar IM, et al. Athletic pubalgia and the "sports hernia": MR imaging findings. Radiology. 2008;247:797–807.
46. Rossi F, Dragoni S. Acute avulsion fractures of the pelvis in adolescent competitive athletes: prevalence, location and sports distribution of 203 cases collected. Skeletal Radiol. 2001;30(3): 127–31.
47. Stevens MA, El-Khoury GY, Kathol MH, et al. Imaging features of avulsion injuries. Radiographics. 1999;19:655–72.
48. Egol KA, Koval KJ, Kummer F, Frankel VH. Stress fractures of the femoral neck. Clin Othop Relat Res. 1998;348:72–8.
49. Warden SJ, Burr DB, Brukner PD. Stress fractures: pathophysiology, epidemiology, and risk factors. Curr Osteoporos Rep. 2006;4(3):103–9.
50. Clement DB, Ammann W, Taunton JE, Lloyd-Smith R, Jesperson D, McKay H, Goldring J, Matheson GO. Exercise-induced stress injuries to the femur. Int J Sports Med. 1993;14(6):347–52.
51. Choi H, McCartney M, Best TM. Treatment of osteitis pubis and osteomyelitis of the pubic symphysis in athletes: a systematic review. Br J Sports Med. 2011;45(1):57–64.
52. Koulouris G. Imaging review of groin pain in elite athletes: an anatomic approach to imaging findings. Am J Roentgenol. 2008;191:962–72.

Part 2

Hip Pathology

5 Femoroacetabular Impingement

Raul Zini and Manlio Panascì

5.1 Introduction

Femoroacetabular impingement (FAI) has been revealed as a significant cause of groin pain and as a predictor of early-onset hip osteoarthritis [1]. This term describes the underlying structural abnormalities in bony morphology, which may alter the force distribution in the joint and can potentially cause injury to the capsulo-labral structure and articular cartilage. FAI describes two main variations of morphologic abnormalities of the hip and resultant observed patterns of chondral and labral injury [2]:

- Cam-type impingement is correlated to an abnormal conformation of the femoral head (Picture 5.1).
- Pincer type is caused by an excessive coverage by the acetabular cavity toward the femoral head (Picture 5.2).

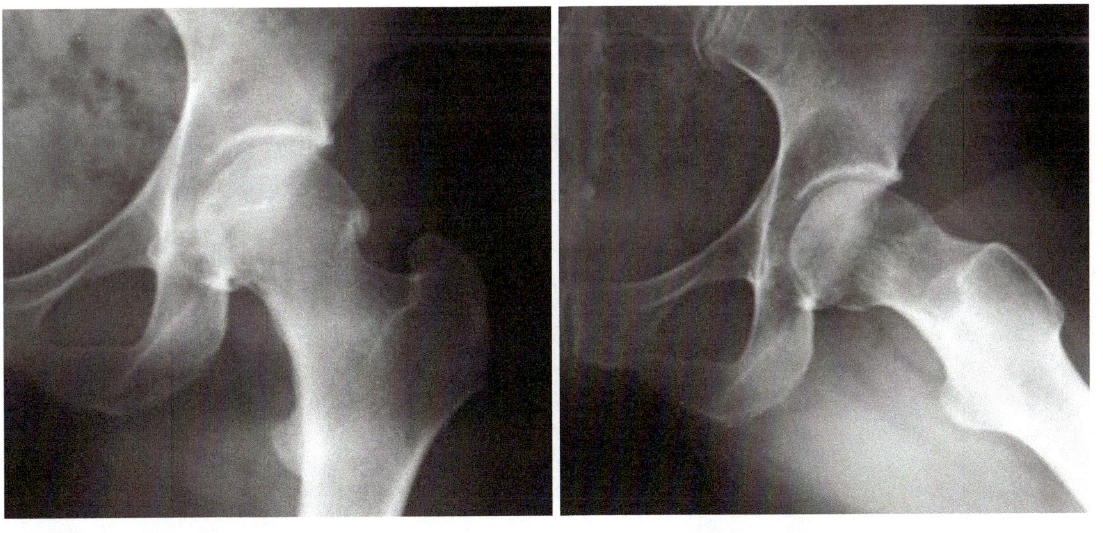

Picture 5.1 Cam impingement

R. Zini, M.D.
Maria Cecilia Hospital, GMV Care and Research, Cotignola, RA, Italy

M. Panascì, M.D. (✉)
San Carlo di Nancy Hospital, GVM Care and Research, Rome, Italy
e-mail: manliopa@gmail.com

R. Zini et al. (eds.), *Groin Pain Syndrome*, DOI 10.1007/978-3-319-41624-3_5

In the mixed type, cam and pincer are contemporarily present in the same hip (Picture 5.3).

Cam: The bump (Picture 5.4) is typically located on the anterolateral aspect of the head–neck junction; it has continuous pressure on the labrum, gradually causing a delamination of the chondral acetabular surface, which is just adjacent to the labrum (Picture 5.5). The micro-traumatic process is due to a tangential force vector created by the bump against the acetabular edge, which finally leads to a cartilage flap delamination with a progressive enlarging area of exposed subchondral bone [3, 4].

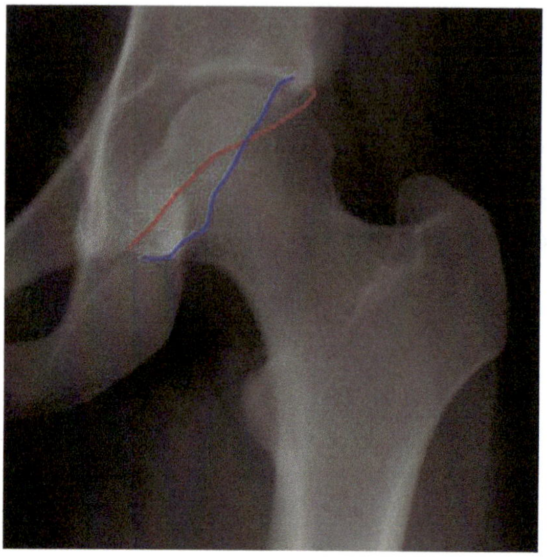

Picture 5.2 Pincer impingement

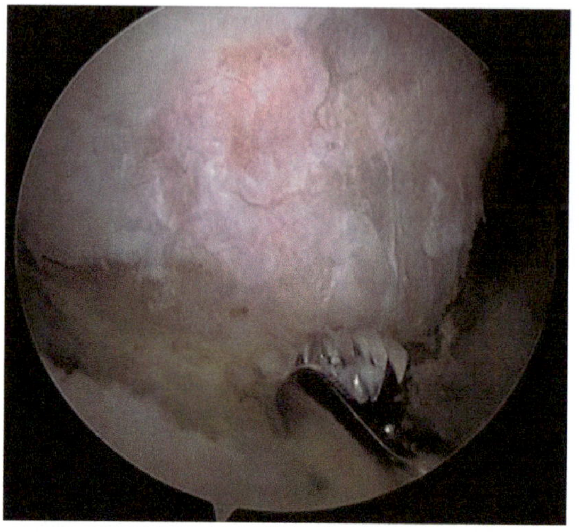

Picture 5.4 Femoral bump

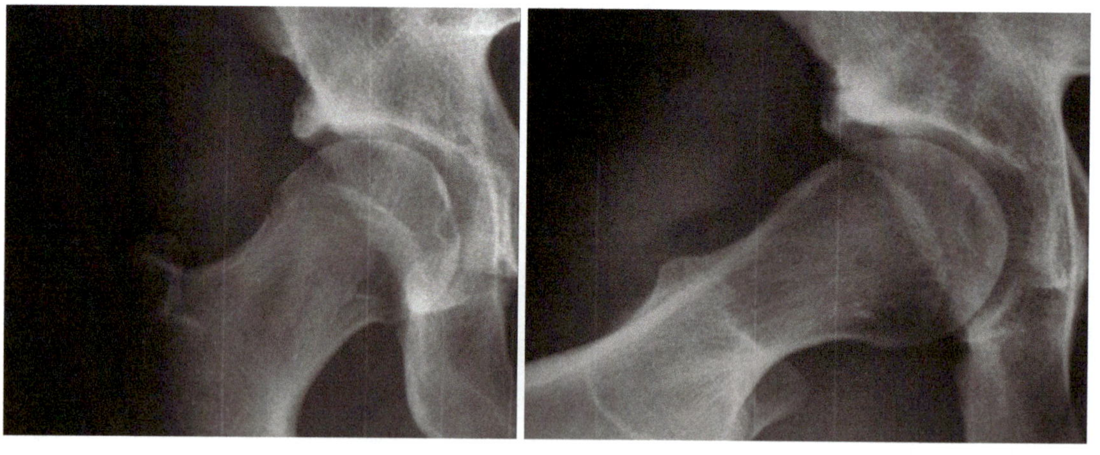

Picture 5.3 Mixed impingement

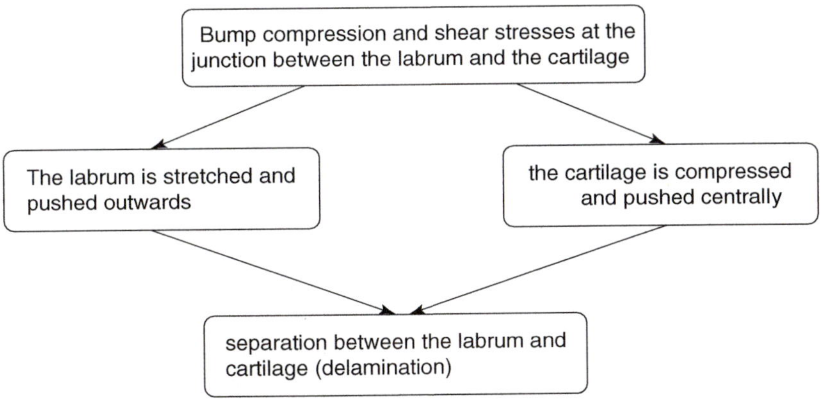

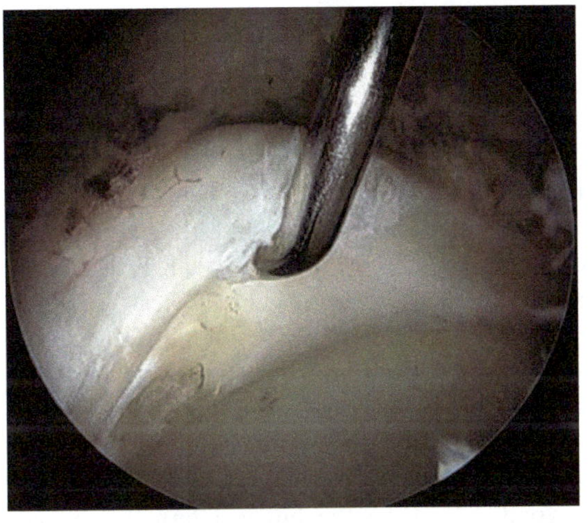

Picture 5.5 Chondrolabral junction

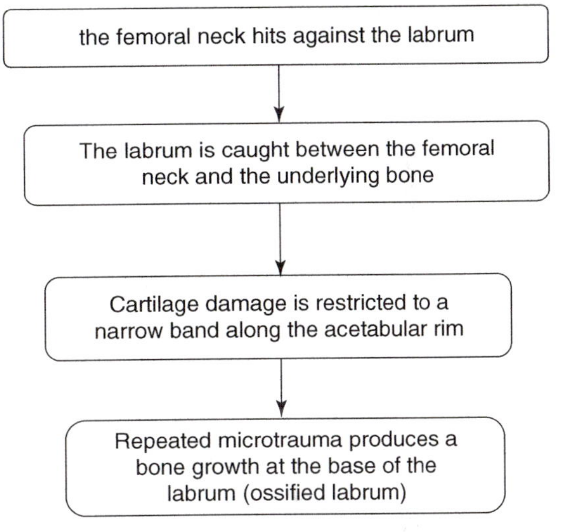

Pincer: The dominant feature is that of a deep socket, in which the range of movement of the hip is limited by the overcovering acetabular rim (Picture 5.6). In coxa profunda, the prototype for pincer impingement, the deep socket limits movement in all directions and leads to a more circumferential pattern of damage [5, 6]. These labral lesions explain the different findings at arthroscopy with more posterior lesions.

Condition resulting from abnormal contact between the femoral head–neck junction and the acetabular rim due to bony abnormalities of the proximal femur (cam type) and/or acetabulum (pincer type) has led to increasing recognition of its role as a cause of groin pain. The prevalence of these hip abnormalities is considerably high; even though radiologic findings do not always correlate with clinic evaluation, the presence of FAI may be a predisposing factor for developing

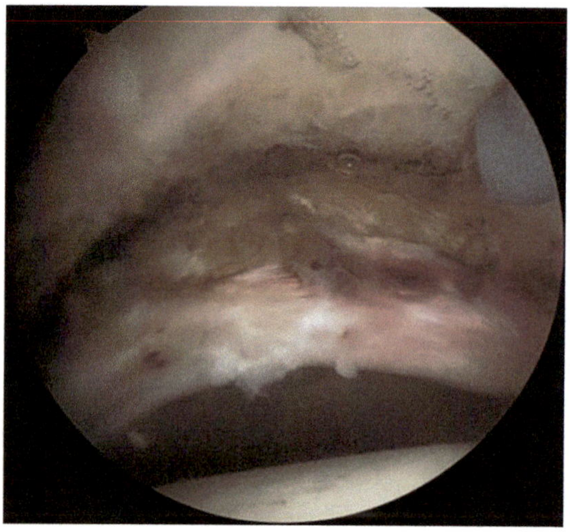

Picture 5.6 Pincer impingement

groin-related sports injuries or indicate a common underlying pathogenic mechanism [7, 8].

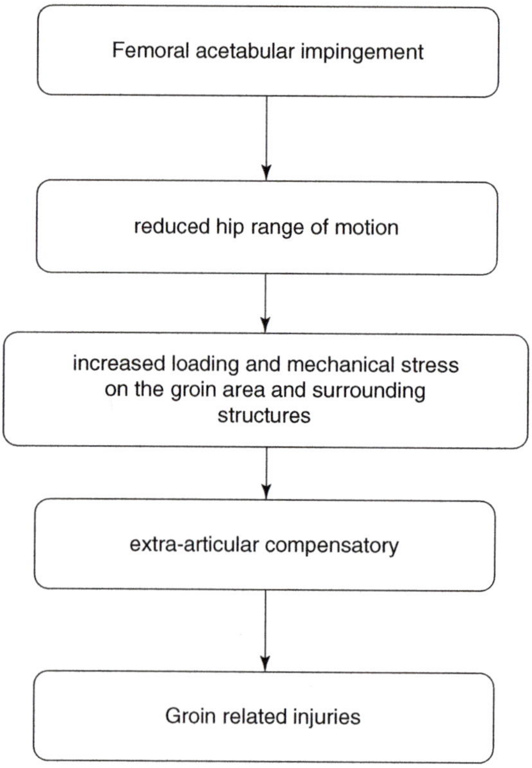

5.2 Clinical Evaluation

The most common pain location in patients with FAI is the groin region, yet many patients had associate discomfort in the lateral hip, thigh, buttock, and low back regions. FAI is commonly manifested as insidious onset of groin pain. As symptoms progress most patients experience moderate to marked pain and substantial limitations of activity.

At the beginning of the symptoms caused by cam impingement, the acetabular cartilage damage is very often already present, but symptoms may be very occasional. In case of pincer impingement, the compression of the acetabular labrum activates the proprioceptive fibers belonging to the labrum itself, so pain precociously appears and patients affected by pincer impingement usually consult an orthopedic surgeon before a cartilage damage has occurred [9].

The majority of patients suffering from chondrolabral pathology associated with FAI show a positive anterior impingement test (*f*lexion *ad*duction, and *i*nternal *r*otation (FADIR) test), which is considered as the most reliable and consistent physical exam finding. In anterior impingement test, the patient is in a supine position and the affected hip is passively flexed beyond 90°. Then the hip is internally rotated while an adduction force is applied. The test is considered positive if reproducing patient typical hip or groin pain.

*I*nternal *r*otation *o*ver *p*ressure (IROP) test is a pain provocation maneuver where the patient is placed in 90° of flexion and end-range internal rotation. A posteriorly directed force is placed through the femur. Reproduction of pain (typically in the groin) indicates a possible intra-articular hip disorder.

The Patrick's or *f*lexion, *ab*duction, and *e*xternal *r*otation (FABER) test has been originally described to differentiate sacroiliac from hip pathologies. The FABER test is often positive in FAI patients and reproduces the anterior hip pain.

In case of SI joint pathology, a posterior pain is referred.

In the resisted straight leg raise test, the hip joint is actively flexed to approximately 30°, with the knee extended. The examiner applies pressure just above the knee, toward the examination table. The test is positive if it reproduces groin pain. Other helpful but less specific and sensitive tests are the log roll test and the posterior impingement test.

These hip provocation maneuvers are a useful part of an evaluation that includes history, further examination findings, and other diagnostic studies. At the moment, none of these diagnostic tests available can reliably confirm or discard the diagnoses of FAI and/or labral pathology in clinical practice in order to support diagnostic imaging and subsequent surgical decision-making. Thus, FAI diagnosis has to be confirmed through an accurate radiologic evaluation, as described in a specific chapter of this book [10, 11].

5.3 Surgical Technique

Supine position is preferred in our routine. Traction is given with a dedicated system to obtain sufficient space and to perform arthroscopy without risk for the intra-articular structures. We routinely use a 20 cm post pad in order to minimize risk of pudendal neurapraxia. Time of traction should always be no more than 2 h [12]. Mild traction is applied to the contralateral limb, which is slightly abducted; this allows for adduction of the operative leg and the C-arm of image intensifier to come in between the legs to obtain anteroposterior (AP), Dunn, and frog-lateral views of the hip.

Standard anterolateral and mid-anterior portals are usually sufficient to perform the entire surgical procedure. We always start with a central compartment evaluation and we routinely perform interportal capsulotomy, which allows a safer and easier instrument movement around the hip joint. Every structure is assessed to look for any pathologic sign, and treatment of every possible causes of pain is achieved. The entire arthroscopic procedure is performed with a 70° arthroscope to identify any labral or acetabular chondral lesions as well as the impingement: In case of a labrum tear, we usually reattach it with resorbable suture anchors; a debridement is a second choice only in case of degenerative lesions. Labral repair with a simple looped stitch, labral base stitch, or a vertical mattress technique is chosen depending on labrum dimensions. It represents an evolution for orthopedic sport medicine, and although reports of labral debridement have been promising, restoring the normal chondrolabral junction with suture anchor repair techniques can potentially provide a more viable option for healing potential of the labrum [13, 14].

Chondral pathologies are treated with shaving, microfractures, or biomimetic scaffold positioning depending on the age of the patient and the size of the defect [15, 16].

Only in cases where instability could be a consequence of our capsulotomy, the medial and lateral capsular edges are anatomically reduced. By use of a suture passer, anatomic side-to-side stitches are placed from distal to proximal until a complete closure is achieved.

5.3.1 Pincer FAI

Pincer-type impingement decompression is achieved under fluoroscopic control: after detaching the labrum from bone with a beaver blade, trimming of the rim with a 5.5 mm round-tip burr is performed; we perform arthroscopic rim trimming typically from the 12 o'clock position down to usually around the 2 to 3 o'clock anterior position (Picture 5.7). Intermittent fluoroscopic imaging is commonly used during the entire procedure. Once the arthroscopic burr tip reaches our desired level of correction, the rim resection is complete

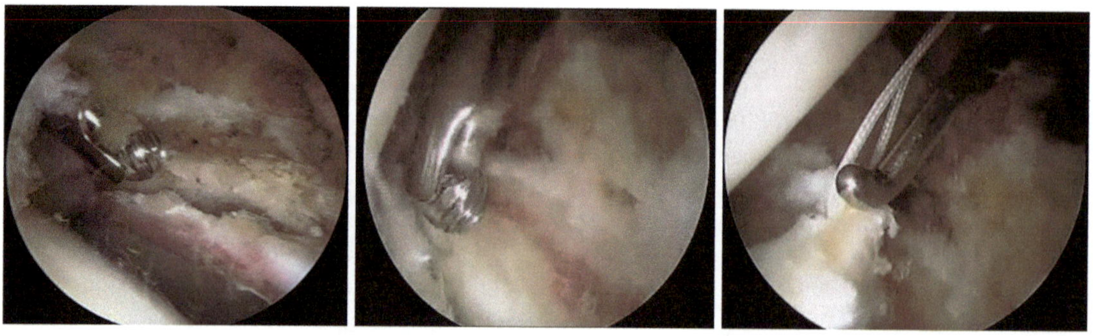

Picture 5.7 From left to right: labrum detachment; rim trimming; suture refixation

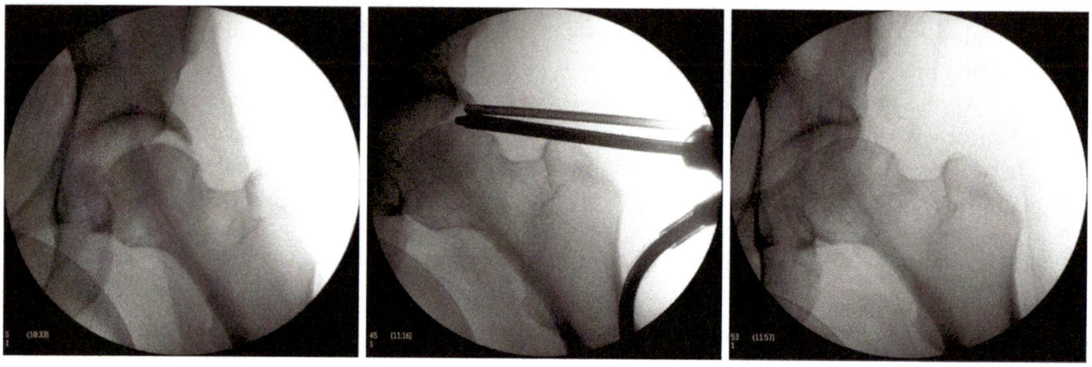

Picture 5.8 Fluoroscopic evaluation during rim trimming

for that level of acetabulum [17, 18] (Picture 5.8). If a sub-spine impingement is present, arthroscopic decompression is achieved. In patients who had showed relief from psoas injections, trans-capsular evaluation of iliopsoas tendon is executed just before removing traction. A correlation between the psoas tendon and the status of the anterior labrum at approximately the 3 o'clock position is performed. In case of fraying or an erythematous, contusion-type lesion of the labrum, we perform fractional release of the tendinous portion of the iliopsoas musculotendinous unit with a radio-frequency probe. A fractional lengthening of the psoas is also performed in case of impingement with the medial portion of the acetabular rim [19, 20].

5.3.2 Cam FAI

Cam-type impingement treatment is possible without traction and a 45° hip flexion. It is sometimes indicated to make a longitudinal capsulotomy to better reach the head–neck junction. Scope is inserted from the anterolateral portal, while the instruments (shaver, burr) are inserted from the mid-anterior portal, but switching is common and very useful for a better three-dimensional assessment (Picture 5.9). Lateral-based lesions are challenging due to the intimate location of the retinacular vessels; thus, proper attention should be given to the vascular anatomy. Osteochondroplasty should include all pathologically appearing cartilage, but shall not go higher or more proximally to the epiphyseal scar, which can be confirmed fluoroscopically [21, 22]. The hip is brought in flexion and internal rotation to directly visualize if a sufficient amount of bone has been removed. Anteroposterior views with the hip in extension and neutral rotation will show the lateral proximal femur, while Dunn view and the frog-lateral position will show the anterior neck aspect [23].

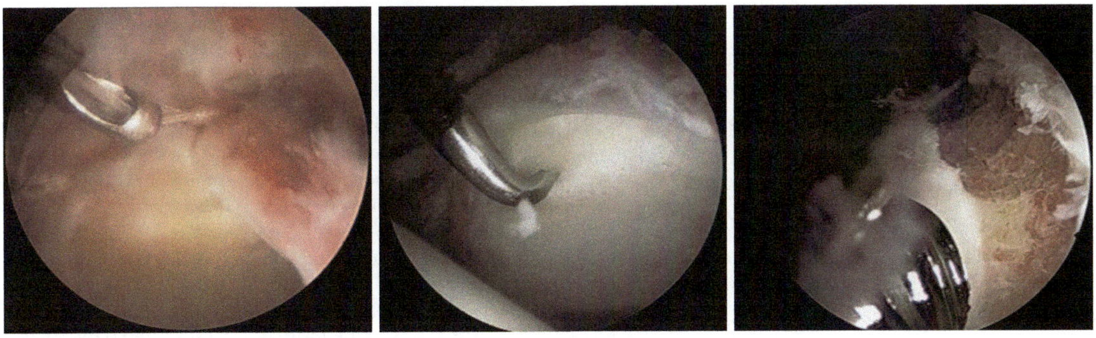

Picture 5.9 From left to right: labral repair; acetabular microfractures; femoroplasty

Table 5.1 Possible complications in hip arthroscopy

Abdominal extravasations
Avascular necrosis of the femoral head
Cartilage scuffings
Hematoma
Instability
Infection
Instrumental breakage
Neurapraxias (femoral, sciatic, perineal)
Over-resection or under-resection of the femoral neck

5.4 Complications

The frequency of complications reported for hip arthroscopy for all indications is generally less than 1.5%, suggesting that the procedure is safe [24]. There is of course a series of specific steps for hip arthroscopy, and the possibility of complications is directly related to those. All possible complications are shown in Table 5.1.

We have reported a 1.5% of complications in our consecutive series of 962 patients. The most common complication is the pudendal neurapraxia (1.2%), which always resolved spontaneously.

5.5 Outcome

Several published articles have been written on outcome after hip arthroscopy.

Results of these studies show that athletes with FAI can return to high-level competitive sport following this procedure [25].

Philippon has published a cohort study of 28 professional hockey players who underwent hip arthroscopy for FAI. The return to sport was 3.8 months (range, 1–5 months) with MHHS of 95 at follow-up. Patients with symptoms lasting less than 1 year returned to sport at 3 months, but patients who delayed surgery over 1 year returned to sport at 4.1 months [26].

Brunner et al. in 2009 reported values return to full sporting activity in 68.8% of cases.

Another study by Nho et al. estimated the return to sport in patients undergoing hip arthroscopy up to 83% [27].

Byrd in 2009 reported its results with a mean follow-up of 27 months. In 90% of professional athletes and 85% of college, there was a return to full sports activity [23].

A recent systematic review showed a high rate of return to pre-injury activity level in athletes treated for FAI. Results achieved a 92% rate return to activity, observed in athletic populations across a variety of sports, with 88% of athletes returning to pre-injury activity levels of participation [28].

In our case series, we have treated since 2009 more than 900 patients with FAI. Among them 65 patients are professional players. Clinical results have shown a return to play rate of 94%, with a mean time recovery of 3.9 months.

Arthroscopic management among athletes is very favorable, but often performed when an important damage has occurred. Substantial secondary damage is frequently present that cannot be completely reversed. We suggest a more comprehensive evaluation including assessment of structural hip anatomy because compensatory hip disorders in the groin area may be due to underlying structural hip disease. In fact FAI is very often not recognized, leading to a delay for a precise diagnosis. Early recognition and treatment have been demonstrated to have a tremendous impact on outcome; Philippon has showed that patients with symptoms lasting less than 1 year returned to sport at 3 months, but patients who delayed surgery over 1 year returned to sport at 4.1 months [29, 30].

Identification of FAI is challenging and symptoms are commonly confused with other hip problems, but many people suffering with groin pain may have femoral acetabular impingement and not realize it. A comprehensive evaluation of all the possible causes of pain in the groin area is mandatory, thus allowing a precise differential diagnosis and a consequent targeted solution.

References

1. Ganz R, Parvizi J, Beck M, et al. Femoroacetabular impingement: a cause for osteoarthritis of the hip. Clin Orthop Relat Res. 2003;417:112–20.
2. Ganz R, Leunig M, Leunig-Ganz K, Harris WH. The etiology of osteoarthritis of the hip: an integrated mechanical concept. Clin Orthop Relat Res. 2008;466:264–72.
3. Johnston TL, Scheker BS, Phillippon MJ, et al. Relationship between offset angle alpha and hip chondral injury in femoroacetabular impingement. Arthroscopy. 2008;24(6):669–75.
4. Gosvig KK, Jacobsen S, Sonne-Holm S, Gebuhr P. The prevalence of cam-type deformity of the hip joint: a survey of 4151 subjects of the Copenhagen Osteoarthritis Study. Acta Radiol. 2008;49:436–41.
5. Steppacher SD, Tannast M, Werlen S, Siebenrock KA. Femoral morphology differs between deficient and excessive acetabular coverage. Clin Orthop Relat Res. 2008;466:782–90.
6. Tanzer M, Noiseux N. Osseus abnormalities and early osteoarthritis: the role of hip impingement. Clin Orthop Relat Res. 2004;428:170–7.
7. Hack K, Di Primio G, Rakhra K, Beaulé PE. Prevalence of cam-type femoroacetabular impingement morphology in asymptomatic volunteers. J Bone Joint Surg Am. 2010;92(14):2436–44.
8. Hammoud S, Bedi A, Magennis E, Meyers WC, Kelly BT. High incidence of athletic pubalgia symptoms in professional athletes with symptomatic femoroacetabular impingement. Arthroscopy. 2012;28(10):1388–95.
9. Lorentzon R, Wedren J, Pietila T. Incidence, nature and cause of ice hockey injuries: a three-year prospective study of a Swedish elite ice hockey team. Am J Sports Med. 1998;16:392–6.
10. Pacheco-Carrillo A, Medina-Porqueres I. Physical examination tests for the diagnosis of femoroacetabular impingement. A systematic review. Phys Ther Sport. 2016;21:87–93.
11. Reiman MP, Mather 3rd RC, Cook CE. Physical examination tests for hip dysfunction and injury. Br J Sports Med. 2015;49(6):357–61.
12. Clarke MT, Villar RN. Hip arthroscopy: complications in 1054 cases. Clin Orthop Relat Res. 2003;406:84–8.
13. Kelly BT, Shapiro GS, Digiovanni CW, Buly RL, Potter HG, Hannafin JA. Vascularity of the hip labrum: a cadaveric investigation. Arthroscopy. 2008;21:3–11.
14. Bedi A, Chen N, Robertson W, Kelly BT. The management of labral tears and femoroacetabular impingement of the hip in young, active patient. Arthroscopy. 2008;24:1135–45.
15. Sampson TG. Arthroscopic treatment for chondral lesions of the hip. Clin Sports Med. 2011;30:331–48.
16. Guanche CA, Sikka RS. Acetabular labral tears with underlying chondromalacia: a possible association with high-level running. Arthroscopy. 2005;21(5):580–5.
17. Gross CE, Salata MJ, Manno K, Yelavarthi V, Barker JU, Williams J, Virkus W, Bush-Joseph C, Nho SJ. New radiographic parameters to describe anterior acetabular rim trimming during hip. Arthroscopy. 2012;28(10):1404–9.
18. Bhatia S, Lee S, Shewman E, Mather 3rd RC, Salata MJ, Bush-Joseph CA, Nho SJ. Effects of acetabular rim trimming on hip joint contact pressures: how much is too much? Am J Sports Med. 2015;43(9):2138–45.
19. Philippon MJ, Devitt BM, Campbell KJ, Michalski MP, Espinoza C, Wijdicks CA, Laprade RF. Anatomic variance of the iliopsoas tendon. Am J Sports Med. 2014;42(4):807–11.
20. Ilizaliturri Jr VM, Buganza-Tepole M, Olivos-Meza A, Acuna M, Acosta-Rodriguez E. Central compartment release versus lesser trochanter release of the iliopsoas tendon for the treatment of internal snapping hip: a comparative study. Arthroscopy. 2014;30(7):790–5.
21. Larson CM, Wulf CA. Intraoperative fluoroscopy for evaluation of bony resection during arthroscopic management of femoroacetabular impingement in the supine position. Arthroscopy. 2009;25(10):1183–92.
22. Lee CB, Clark J. Fluoroscopic demonstration of femoroacetabular impingement during hip arthroscopy. Arthroscopy. 2011;27(7):994.
23. Byrd JW, Jones KS. Arthroscopic femoroplasty in the management of cam-type femoroacetabular impingement. Clin Orthop Relat Res. 2009;467(3):739–46.
24. Ilizaliturri Jr VM. Complications of arthroscopic femoroacetabular impingement treatment: a review. Clin Orthop Relat Res. 2009;467(3):760–8.
25. Philippon MJ, Arnoczky SP, Torrie A. Arthroscopic repair of the acetabular labrum: a histologic assessment of healing in anovine model. Arthroscopy. 2007;23:376–80.

26. Philippon MJ, Wiess DR, Kuppersmith DA, Briggs KK, Hay CJ. Arthroscopic labral repair and treatment of femoroacetabular impingement in professional hockey players. Am J Sports Med. 2010;38(1):99–104.
27. Brunner A, Horisberger M, Herzog RF. Sports and recreation activity of patients with femoroacetabular impingement before and after arthroscopic osteoplasty. Am J Sports Med. 2009;37(5):917–22.
28. Alradwan H, Philippon MJ, Farrokhyar F, et al. Return to preinjury levels after surgical management of femoroacetabular impingement in athletes. Arthroscopy. 2012;28(10):1567–76.
29. Philippon M, et al. Femoroacetabular impingement in 45 professional athletes: associated pathologies and return to sport following arthroscopic decompression. Knee Surg Sports Traumatol Arthrosc. 2007;15(7):908–14.
30. Malviya A, Paliobeis CP, Villar RN. Do professional athletes perform better than recreational athletes after arthroscopy for femoroacetabular impingement? Clin Orthop Relat Res. 2013;471(8): 2477–83.

6 Acetabular Labral Tear

Christoph Gebhart

6.1 Introduction

The acetabular labrum (AL) plays an integral role in the hip joint with many functions, including shock absorption, joint lubrication, pressure distribution, and aiding in hip stability [1]. Damage to the AL may lead to progression of hip osteoarthritis. Labral tears have been associated with femoroacetabular impingement, instability, sports-related activities, trauma, capsular laxity or hypermobility, and dysplasia.

6.2 Anatomy, Physiology

With a broad basis connected to the acetabular rim bone, the AL consists of a circumferential type I collagen fibrocartilaginous ring with a triangular cross section approximately 4.7 mm wide at the bony attachment by approximately 5.5 mm tall [Fig. 6.1, 2, 3]. The anterior collagen fibers of the AL are, different to posterior, parallel to the bony edge and therefore easy to be separated by shear forces. The posterior AL fibers attach perpendicular and merge with the collagen fibers of the bony edge and are therefore more resistant to shear forces [4]. The AL outside is faced to the inside of the hip joint capsule merging on the capsular side with the bony acetabulum and on the articular side with the acetabular hyaline cartilage [5]. The inner layer of the AL on the femoral side is formed as a result of compressive and shear forces. The outer AL layer of dense connective tissue on the capsular side is formed under tension stress [6].

The labrum inside directly rests on the cartilage of the femur head. The intersecting width and thickness measuring lines divide the labrum into inner and outer as well as basic and peripheral quadrants. In these quadrants, labrum alterations can be precisely located and described. An intact acetabular labrum ring is required for normal hip joint biomechanics and cartilage preservation. The AL also known as the cotyloid ligament and the transverse acetabular ligament create a circular seal around the femur head that maintains hip joint pressurization. The transverse acetabular ligament is fixed to the anterior and posterior pillar of the acetabular notch and origin of the ligamentum capitis femoris. AL and transverse ligament are encompassing more than 50% of the femoral head going lateral to the femur equator and thereby stabilizing the femoral head during range of motion [1]. Luxation of the femur head results in negative pressure inside the acetabular ring, and therefore this labrum seal mechanism contributes to antiluxating hip stability. Disruption of the labrum increases the rate of acetabular cartilage compression and the contact stress between the femoral

C. Gebhart, M.D.
Vienna International Center, Private Clinic Döbling, Center for Minimal Invasive Hip Surgery, Heiligenstätterstrassse 57-63, Vienna 1190, Austria
e-mail: president@asha.at

R. Zini et al. (eds.), *Groin Pain Syndrome*, DOI 10.1007/978-3-319-41624-3_6

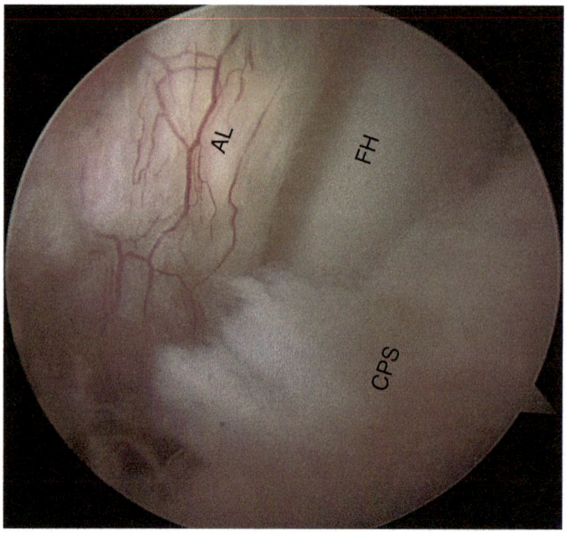

Fig. 6.1 Intact acetabular labrum

and acetabular cartilage. Henak et al. [7] suggest that the labrum with 4–11% of the total load transferred across the joint plays a larger role in load transfer and joint stability in hips with acetabular dysplasia. In hips with normal acetabular geometry, the AL supported only 1–2%. Despite the increased load transferred to the acetabular cartilage in simulations without the labrum, there were minimal differences in cartilage contact stresses. This was because the load supported by the cartilage correlated to the cartilage contact area. A higher percentage of load was transferred to the labrum in the dysplastic model because the femoral head achieved equilibrium near the lateral edge of the acetabulum. Comparable to the fibrocartilaginous meniscus ring of the knee joint, the vascularization of the AL shows different zones of blood supply. The AL blood supply arises from an artery retinaculum formed by the superior and inferior gluteal arteries as well as the medial and lateral circumflex arteries. The majority of vessels enters near the labrum basis—compared to the meniscal red zone—providing a network of supporting arteries at the labrum outside. Therefore, the AL receives its blood supply from radial branches of a periacetabular periosteal vascular ring that traverses the osseolabral junction on its capsular side and continues toward the labrum's free edge. The hip capsule, the synovial lining, and vessels from the osseous acetabular rim do not appear to provide substantial contributions to the labral blood supply [8]. Many different types of free sensory nerve endings and sensory nerve end organs mostly in the AL outer layer including Vater–Pacini (pressure), Golgi–Mazzoni (pressure), Ruffini (deep sensation, temperature), and articular (Krause) corpuscles (pressure, pain) transmit pain, tactile sensation, and temperature [9]. Between the labrum outside and the inside of the adjoining hip capsule, usually a perilabral sulcus guarantees free inter-movability of both structures. At the joint side the labrum is usually directly connected to the hyaline cartilage of the acetabulum in a smooth surface. This junction is called chondrolabral transition zone. Sometimes a small gap persists physiological maturing and leaves an incomplete connected transition zone called sublabral sulcus [Fig. 6.2]. A sublabral sulcus therefore is not classified as a labrum rupture and might have no pathological meaning. If it is a predisposition for labrum rupturing or improper function, it has to be proved in further studies. Kappe et al. [10] found that in the presence of impingement-inducing deformity, the extent of deformation is not associated with the incidence of labral lesions. The severity of labral lesions correlated to the severity of acetabular chondromalacia as well as patient age. Labral lesions are associated with early degenerative hip disease in FAI.

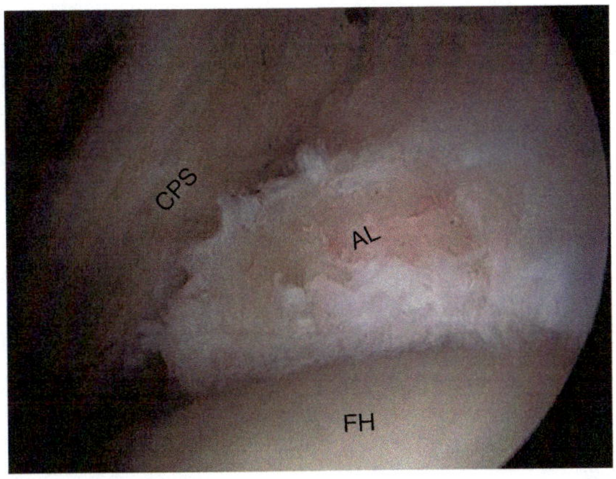

Fig. 6.2 Ossified acetabular labrum

6.3 Labral Tear Diagnosis: Rupture Classification

Labral tears frequently go undiagnosed during an extended period of time because of the vast differential diagnosis and the need for specialized diagnostic tools.

A prevalence of 22–55% for labral tears has been reported in patients with hip or groin pain [8]. Peterson published the first acetabular labrum tear in 1957 and described two cases of labral tears associated with irreducible posterior hip dislocation. Several types of labrum alterations from fatty degeneration to intra-articular cyst formations, ossifications, or fibrillations have been described. To select the appropriate treatment, it is essential to differ between these variable labrum alterations and know about their natural development and progression. Evaluation usually begins with plain radiographs to assess for dysplasia, degeneration, and other causes of pain. The preoperative assessment of labral tears has to be done by clinical examination and analysis of hip MRI findings. The typical clinical picture and most consistent physical examination finding presents anterior hip or groin pain and positive anterior impingement tests, eventually aggravated by activity. Sometimes there are also mechanical symptoms including locking, clicking, or giving way.

Because the majority of labral tears is located anterior/superior, most frequently impingement tests reveal inguinal pain at hip flexion + internal/external rotation. Trauma anamnesis with dorsal (sub)luxations can lead to more posterior tears representing dorsal, gluteal, and less commonly buttock or sciatic referred pain.

Magnetic resonance arthrography (MRA) is sensitive and accurate for the detection of acetabular labral tears. Several studies acknowledge excellent intraobserver sensitivity, specificity, and reliability to the MRI findings of the still most commonly used Czerny labral tear classification. The Czerny MRA classification system differs between stage 0 (normal labrum), stage I (1A, increased signal intensity within labrum; 1B, labrum thickened without perilabral recessus), stage II (2A, extended intrasubstance tears; 2B, thickened without perilabral recessus), and stage III (A, complete avulsion; B, complete avulsion + deformation) acetabular labral tears.

A clockface description provides the best way to accurately localize a labral tear and define its extent [11–13].

An intraoperative AL rupture assessment classification system published by Lage et al. [14] underlies the patterns of labral tears found in 37 patients undergoing hip arthroscopy. The classification categories found were in terms of etiology (traumatic, 18.9% of cases; degenerative, 48.6%; idiopathic, 27.1%; congenital, 5.4%) and morphology (radial flap, 56.8%; radial fibrillated, 21.6%; longitudinal peripheral, 16.2%; unstable, 5.4%).

The classification system published by Beck et al. [11] differs between normal, degenerated, full-thickness tears, detachment, and ossification of AL alterations.

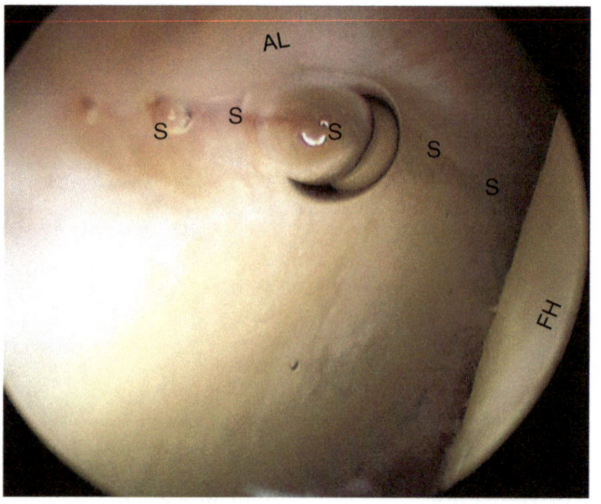

Fig. 6.3 Intact acetabular labrum at the sublabral sulcus transition zone

Because hip arthroscopy outcome research is limited by the lack of an accurate, universally used terminology to describe the characteristics and location of intra-articular lesions, a more recent classification system was published by the Multicenter Arthroscopy of the Hip Outcomes Research Network (MAHORN) group in 2010 [3]. The MAHORN proposed classification of labral tears differs between normal, hypoplastic/hyperplastic tears, complex/degenerative, labral–chondral separation, and partial, complete, or flap AL ruptures. According intrasubstance AL changes, they suggest to distinguish mucinoid/yellow, floppy, bruising, ossified, or calcific AL alterations.

Summarizing the classification systems differs between different types of labrum ruptures. Various types of AL ossifications can lead to so-called pincer deformities causing hip impingement, pain, and reduction of range of hip joint motion [Fig. 6.3].

6.4 Acetabular Labral Tears: Treatment

Treatment begins usually conservatively with rest and nonsteroid anti-inflammatory drugs. The benefit of physical therapy is discussed controversially. At therapy, resistant painful cases with ongoing hip joint degeneration surgical treatment becomes necessary including arthroscopic debridement or suturing of labral tears with repair of underlying or associated structural problems.

Literature supports repairing of unstable labral tears of sufficiently good tissue quality to preserve the physiological function. In an ovine model, fibrovascular scar healing to the acetabular bone of a debrided labrum has been shown after single suture anchor repair [15]. A routine fixation of all degenerative labral tears is not supported [16].

In principle, there are options of open or arthroscopical surgical procedures. Either (partial) debridement or refixation or even transplantations of the AL can be performed. Arthroscopical labral repair has been reported to improve clinical outcomes with lesser pain and faster recovery compared to open surgery.

Philippon et al. [3, 15, 17] showed arthroscopic repair of labral pathology instead of debridement to be successful for improvement of symptoms and function with satisfying results and good outcomes at 2-year follow-up performed at preoperative joint space greater than 2 mm. For the detached or torn AL with adequate tissue (>7 mm), labrum repair is recommended.

When the labrum basis remains stable and causes neither pain nor impingement despite a persistent gap of the chondrolabral transition zone, there is no indication for labral refixation. At bruised AL, a rim reduction and labrum repair shall be performed and the chondrolabral transition zone restored in cases of cartilage damage.

If applicable, a concurrent cam deformity with underlying or interfering femoral impingement has to be treated additionally with cam resection. In dysplastic cases, the rim reduction has to be evaluated exactly in order to avoid the creation of unstable hip joints with pathological center edge angles of the acetabulum.

Debridement of the AL is best done in a way with preserving enough substance to maintain labrum function. Therefore, some cases of AL ossification resection *FIG* might leave inadequate labrum substance for refixation. For these cases with inadequate remaining labrum substance, successful transplantations of the AL either with tensor fasciae latae, gracilis, or hamstring tendon autografts or allografts have been reported. Anyhow, the AL has to be fixed in a way that provides adequate AL seal with the femoral head.

References

1. Grant AD, Sala DA, Davidovitch RI. The labrum: structure, function, and injury with femoro-acetabular impingement. J Child Orthop. 2012;6(5):357–72. PMCID: PMC3468736.
2. Lee S, Wuerz TH, Shewman E, McCormick FM, Salata MJ, Philippon MJ, Nho SJ. Labral reconstruction with iliotibial band autografts and semitendinosus allografts improves hip joint contact area and contact pressure: an in vitro analysis. Am J Sports Med. 2015;43(1): 98–104. doi:10.1177/0363546514553089. Epub 2014 Oct 31.
3. Philippon MJ, Briggs KK, Yen YM, Kuppersmith DA. Outcomes following hip arthroscopy for femoroacetabular impingement with associated chondrolabral dysfunction: minimum two-year follow-up. J Bone Joint Surg Br. 2009;91(1):16–23. doi:10.1302/0301-620X.91B1.21329.
4. Cashin M, Uhthoff H, O'Neill M, Beaulé PE. Embryology of the acetabular labral-chondral complex. J Bone Joint Surg Br. 2008;90:1019–24. doi:10.1302/0301-620X.90B8.20161.
5. Seldes RM, Tan V, Hunt J, Katz M, Winiarsky R, Fitzgerald Jr RH. Anatomy, histologic features, and vascularity of the adult acetabular labrum. Clin Orthop Relat Res. 2001;382:232–40.
6. Petersen W, Petersen F, Tillmann B. Structure and vascularization of the acetabular labrum with regard to the pathogenesis and healing of labral lesions. Arch Orthop Trauma Surg. 2003;123:283–8.
7. Henak CR, Ellis BJ, Harris MD, Anderson AE, Peters CL, Weiss JA. Role of the acetabular labrum in load support across the hip joint. J Biomech. 2011;44(12):2201–6. doi:10.1016/j.jbiomech.2011.06.011. Epub 2011 Jul 14.
8. Kalhor M, Horowitz K, Beck M, Nazparvar B, Ganz R. Vascular supply to the acetabular labrum. J Bone Joint Surg Am. 2010;92(15):2570–5. doi:10.2106/JBJS.I.01719.
9. Kim YT, Azuma H. The nerve endings of the acetabular labrum. Clin Orthop Relat Res. 1995;320:176–81.
10. Kappe T, Kocak T, Bieger R, Reichel H, Fraitzl CR. Radiographic risk factors for labral lesions in femoroacetabular impingement. Clin Orthop Relat Res. 2011;469(11):3241–7. doi:10.1007/s11999-011-1978-8. Epub 2011 Jul 12.
11. Beck M, Kalhor M, Leunig M, Ganz R. Hip morphology influences the pattern of damage to the acetabular cartilage: femoroacetabular impingement as a cause of early osteoarthritis of the hip. J Bone Joint Surg Br. 2005;87(7):1012–8.
12. Czerny C, Hofmann S, Neuhold A, et al. Lesions of the acetabular labrum: accuracy of MR imaging and MR arthrography in detection and staging. Radiology. 1996;200(1):225–30.
13. Dameron TB. Bucket handle tear of acetabular labrum accompanying posterior dislocation of the hip. J Bone Joint Surg. 1959;41A:131–4.
14. Lage LA, Patel JV, Villar RN. The acetabular labral tear: an arthroscopic classification. Arthroscopy. 1996;12(3):269–72.
15. Philippon MJ, Arnoczky SP, Torrie A. Arthroscopic repair of the acetabular labrum: a histologic assessment of healing in an ovine model. Arthroscopy. 2007;23(4):376–80.
16. Haddad B, Konan S, Haddad FS. Debridement versus re-attachment of acetabular labral tears: a review of the literature and quantitative analysis. Bone Joint J. 2014;96-B(1):24–30. doi:10.1302/0301-620X.96B1.32425.
17. Philippon MJ, Briggs KK, Fagrelius T, Patterson D. Labral refixation: current techniques and indications. HSS J. 2012;8(3):240–4.

7 Pathology of Central Compartment

Nicolas Bonin and Antoine Dangin

7.1 Synovial Pathologies

Synovial pathologies should be considered when there is a mechanical or inflammatory, unilateral and progressive joint pain. Complains of blockages or pseudo-blockages must also attract the surgeon's attention to discuss a synovial etiology. X-rays can be useful by highlighting indirect signs such as acetabular or femoral head subchondral cyst and a preserved joint space. MRI with gadolinium vascular injection can show a larger joint space, effusion and pathological contrast-enhanced synovial. However an arthro-CT or an arthro-MRI is often required to describe precisely different pathological synovial appearance and guide towards a diagnostic hypothesis. Finally, only biopsies will give a definitive diagnostic.

The first indications for hip arthroscopy were synovial pathologies. Arthroscopy will allow a macroscopic evaluation of the synovium, a diagnostic confirmation by histological biopsies and the surgical treatment. It mostly consists in partial or subtotal synovectomy and removal of foreign bodies if needed.

Endoscopic appearance of inflammatory arthritis should be well known in order to practice synovial biopsies whenever there is a suspicious appearance of the synovium.

7.1.1 Synovial Chondromatosis

Synovial chondromatosis is the most frequent synovial pathology. The symptoms presented by the patient are joint pain, stiffness and/or blockages. The most affected joints are the hip, the knee and the ankle. Its physiopathology is an abnormal production of joint fluid and chondroid foreign bodies that can calcify. These chondroid bodies can be missed either by an X-ray if they are not calcified or by a standard MRI since their magnetic signal is near the one of joint effusion. Arthro-MRI or arthro-CT are the best radiological imaging exams for cases of chondromatosis (Fig. 7.1). In addition to its diagnosis role, it allows a technical approach for the surgical treatment: it will allow localizing and counting foreign bodies, highlighting the pathological synovial that appears thickened and irregular. Chondromas can be free, pedicle or within the synovium. These three types can be present at the same time [1].

Anatomic evolution is based on three stages, described by Milgram [2]:

N. Bonin, M.D. (✉)
Hip and Sports Medicine Department, Lyon-Ortho-Clinic, 29 Avenue des sources (Bat B), Lyon 69009, France
e-mail: dr.bonin@lyon-ortho-clinic.com

A. Dangin
Service de Chirurgie Orthopédique et de Traumatologie, Centre Hospitalier Universitaire Nord Saint Etienne, Saint Priest en Jarez 42270, France
e-mail: antoine.dangin@gmail.com

R. Zini et al. (eds.), *Groin Pain Syndrome*, DOI 10.1007/978-3-319-41624-3_7

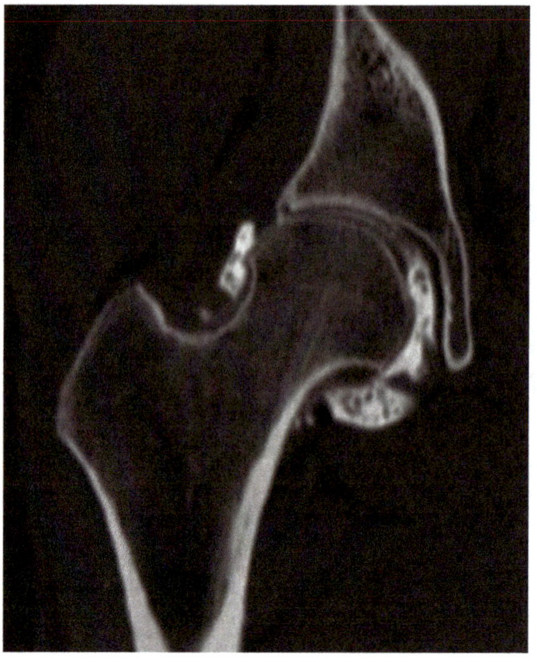

Fig. 7.1 Arthro-CT coronal view of chondroid bodies in a right hip

- Stage 1: intra-synovium chondromas
- Stage 2: intra-synovium and free chondromas
- Stage 3: free chondromas

Concerning the first two stages, the synovium is still active, with a risk of recidivism after treatment. During the last stage, production of chondromas runs out gradually and a treatment will allow healing in most cases.

To reduce morbidity, the treatment can be realized under arthroscopic assistance. It consists in removing all foreign bodies with a synovectomy adapted to the pathology extension.

Arthroscopy will appreciate the synovial status, witness of the evolutionary stage of the illness. Biopsies will allow determining the diagnosis but also determining the activity level of the synovial chondromatosis. An active synovial chondromatosis is determined by the importance of chondroid metaplasia foci (Fig. 7.2). Between each foci, the synovium is normal, hence the importance of realizing multiple biopsies led by arthroscopy.

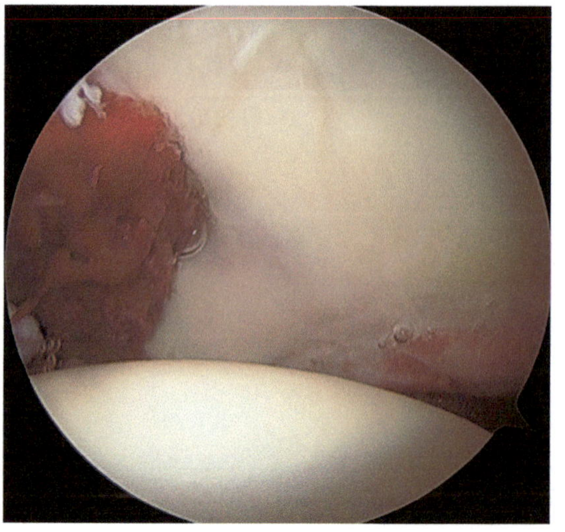

Fig. 7.2 Arthroscopic view of chondroid metaplasia foci of the acetabular fossa in a right hip

Arthroscopic treatment of synovial chondromatosis can be considered as the reference treatment. According to the literature [1, 3, 4], it showed good to excellent results in 48–57% of the cases, with a conversion to total hip replacement (THR) in 17% at 6 years mean follow-up. However, the risk of recurrence remains high (16.2% re-arthroscopy).

Open surgery for total synovectomy shows lower recurrence rate but a higher morbidity and a higher THR conversion rate.

The extended monitoring of a synovial chondromatosis is justified by the possibility of late sarcomatous transformation. The patient should be warned that late evolution is towards secondary osteoarthritis.

Other forms of synovial chondromatosis exists [5]:

- Decalcifying form: regional radiological decalcification of the femoral head, the femoral neck and the acetabulum. The difference with algoneurodystrophy is its longer evolution, the absence of radiological blurring bone contours, little or no scintigraphic uptake and no regional bone oedema on MRI.
- Retractable form: mimicking capsulitis. The joint stiffness is severe and often rebel.

Chondromas are often small and within the synovium or in the rear bottom of the acetabulum.

- Macromedia form or erosive pseudocystic form: one or more broad-rimmed gaps of the femoral neck near synovial reflections, which can mimic a bone cyst. These erosions are related to the pressure on the bone induced by metaplasia of synovial masses in little distensible capsular areas.
- Ossified synovial form: stiffness and almost always ossified. Its functional prognosis is poor.

7.1.2 Villonodular Synovitis

Villonodular Synovitis (VNS) is usually diagnosed next to mechanical pain and blockage complains from the young adult. It is mostly a mono-articular synovitis affecting in the first position, the knee, and in the second position, the hip. MRI aspect of VNS is pathognomonic due to the presence of hemosiderin deposition with ferromagnetic properties giving a particular MRI characteristic (Fig. 7.3).

VNS can occur in two different histopathological presentations: *the diffused form and the localized form.*

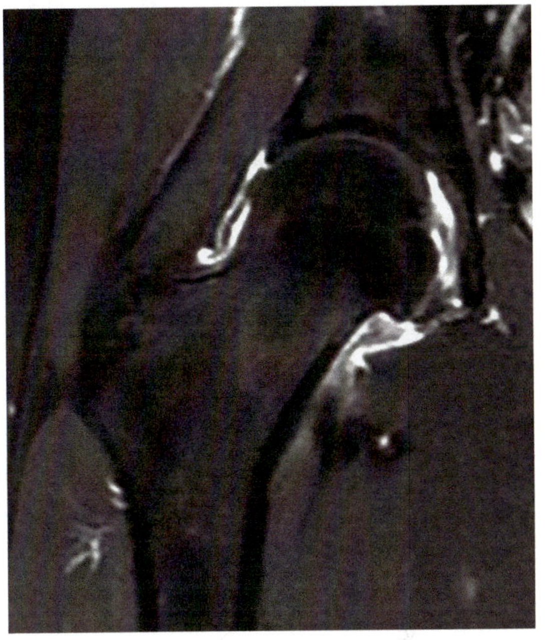

Fig. 7.3 Pathognomonic hemosiderin deposition signal of a villonodular synovitis in T1-FS gadolinium MRI coronal view of the right hip

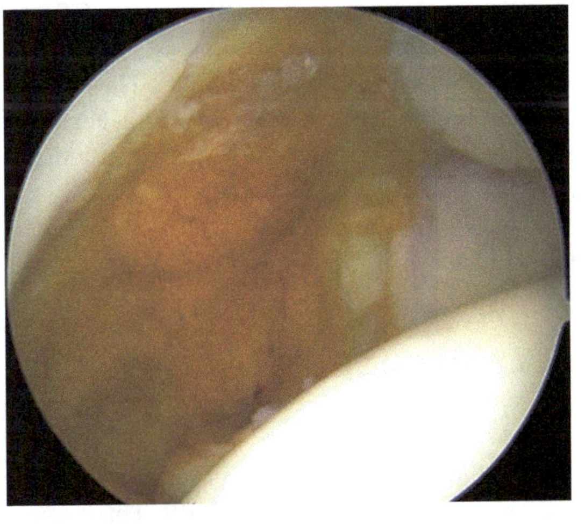

Fig. 7.4 Arthroscopic view of the rust colour synovium of the acetabular fossa of a right hip in the case of a villonodular synovitis

- *The diffused form of VNS* is characterized by a partial or total pathological synovial. It is often associated to a sero-hematic hydrarthrosis. Villi are brown or rust colour due to the presence of hemosiderin deposit (Fig. 7.4). Treatment can be surgical synovectomy with hip dislocation or less invasive by arthroscopic synovectomy. During treatment by arthroscopy, each articular area must be methodically explored and cleaned to be as complete as possible. The main difficulty of a complete arthroscopic synovectomy is to access the posterior compartment. It is increased by the presence of blood vessels tangential to the posterior capsule inducing a high bleeding risk. Therefore, arthroscopic synovectomy cannot be total. It is only justified by its lower aggressiveness in a disease where recidivism after hip dislocation surgery is frequent (up to 50%), leading to osteoarthritis [6, 7]. Recidivism mostly occurs during the first 4 years, sometimes later, justifying an MRI monitoring. In case of recidivism, an early iterative synovectomy must be accomplished in front of the aggressiveness of the disease.

– *Localized form of VNS* is characterized by an intra-articular pedunculated mass. It can become necrotic after a twist mechanism and engendering a mono-arthritis symptomatology. This form of VNS needs its complete removal in a healthy area with a large excision of the pedicle to obtain healing with no risk of recidivism [6]. This can be realized under arthroscopy.

  It is important to keep in memory that an untreated localized form of VNS can evolve to a diffused form.

7.1.3 Mechanical or Inflammatory Synovitis

During an osteoarthritic push, a significant mechanical or inflammatory synovial proliferation is often observed. Voluminous fringes can be trapped between the articular surfaces explaining phenomena of blockages [8]. Arthro-CT or arthro-MRI can show villi of important size that can be confused with foreign bodies. The benefit of arthroscopic synovectomy in this mechanical proliferation has not been proven, and corticosteroid infiltration should be preferred. However, synovial biopsy should be performed whenever the villi have a suspicious aspect.

7.1.4 Medial Synovial Fold Pathology

Some painful hips have been described in association with a synovial fold back [9]. This anomaly concerns the physiological structure called the medial fold or can be pathological and so named as a plicae. Such pathologies involve only a few cases, and it is necessary to stay prudent concerning this etiology for hip pain. Nevertheless, observing such anomalies during hip arthroscopy must question the surgeon on the painful potential to know when to realize its resection.

The anomaly of the medial fold is probably a conflict between itself and the iliopsoas tendon conducting to the development of a mechanical synovitis. The patient usually only complains of a widespread pain. The diagnosis cannot be established by the clinic or imaging. Only arthroscopy can observe the inflammatory area facing a thickened medial fold. The arthroscopic treatment consists in its resection with a shaver or an electrocautery device in hip flexion.

7.2 Septical Arthritis

Septical arthritis diagnostic is often established preoperatively by history, physical exam, biology, radiologic studies and, if any doubts, hip aspiration for microbiology.

Arthroscopic lavage and debridement is actually the recommended treatment for a septical hip arthritis since it is as efficient as opened surgery with lower morbidities [10, 11]. During arthroscopic treatment, multiple bacterial and anatomopathological samples are realized in the suspicious areas. The joint washing must concern the peripheral and central compartment. The excision of all suspicious septical soft tissues or synovial must be completely done with a shaver or an electrocautery device.

As soon as the samples are done, an empiric antibiotic therapy is undertaken with a multidisciplinary agreement [11].

Opened surgery is actually reserved to abscess lesions or to osteomyelitis.

7.3 Foreign Bodies

Cartilage or synovial origin intra-articular foreign bodies can become symptomatic. The first etiology of intra-articular chondral or osteochondral foreign bodies is synovial chondromatosis. But chondral or osteochondral fragments can also come from femoroacetabular impingement, hip trauma or osteoarthritis [12, 13]. They can be size, shape or different volume. Their numbers can be variable. It is necessary to have a proper radiological assessment such as an arthro-CT or an arthro-MRI to count, measure and localize the chondromas, osteochondromas or any other foreign bodies.

Foreign bodies can be removed under arthroscopy (Fig. 7.5). If their size is too important, they need to be fragmented prior to be removed (Fig. 7.6). Sometimes, open surgery may be necessary if the foreign bodies are too big or not fragmentable.

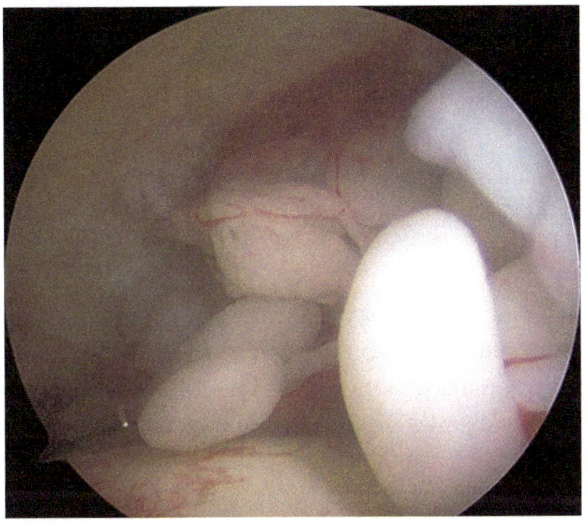

Fig. 7.5 Arthroscopic removal of numerous chondral foreign bodies of the central compartment of the hip in synovial chondromatosis

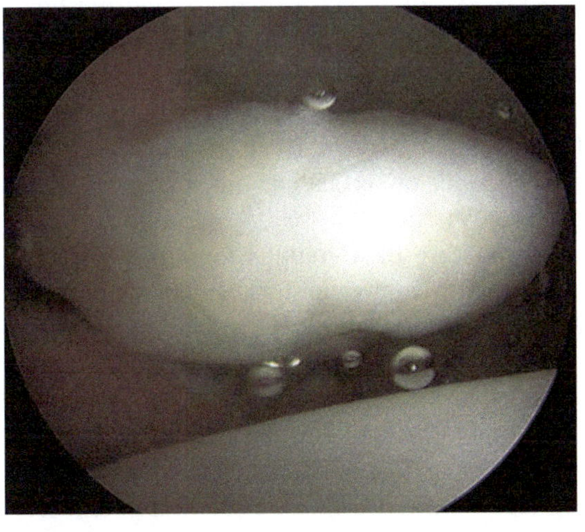

Fig. 7.6 Arthroscopic view of large osteochondral foreign body of the central compartment of a right hip that needs to be fragmented before extraction

7.4 Ligamentum Teres Pathology

7.4.1 Pathological Plicae

Some cases of pathological painful plicae have been published [14]. The plicae were located in the acetabular fovea. Arthro-MRI can highlight a plicae by visualizing a 4–6 mm large strip between the ligamentum teres and the acetabular notch, looking as a duplication of the ligamentum teres.

Arthroscopic resection is carried out under traction, with an articulated electrocautery device introduced by the mid-anterior portal, arthroscopic control being in the anterolateral portal.

7.4.2 Ligamentum Teres Trauma

Traumas of the ligamentum teres (LT) are rare: according to the literature, 5–17.5% of patients undergoing hip arthroscopy present an LT tear [15]. LT tear occurs next to an extension or abduction trauma hip movement.

Patients presenting LT tears complain of groin pain, blockages and instability of the hip joint. These symptoms are not specific to LT tears, which make the diagnostic difficult. A clinical test has been recently described consisting in applying internal and external rotation with a flexed hip up to 70° and 30° abduction. The test is positive when pain is induced by this manoeuvre [16].

X-rays are generally normal but can occasionally show a wider joint space. MRI is capable of diagnosing LT tears with high sensitivity (91%) [17]. Concerning complete LT tears, there has been no diagnosing difference between MRI and arthro-MRI. However, arthro-MRI is more efficient concerning partial and degenerative LT tears [18].

Treatment can be ligamentum teres resection under arthroscopy (Fig. 7.7) with good outcomes [19]. An arthroscopic reconstruction can also be realized [15]: under arthroscopic visualization and fluoroscopic control, a femoral tunnel is

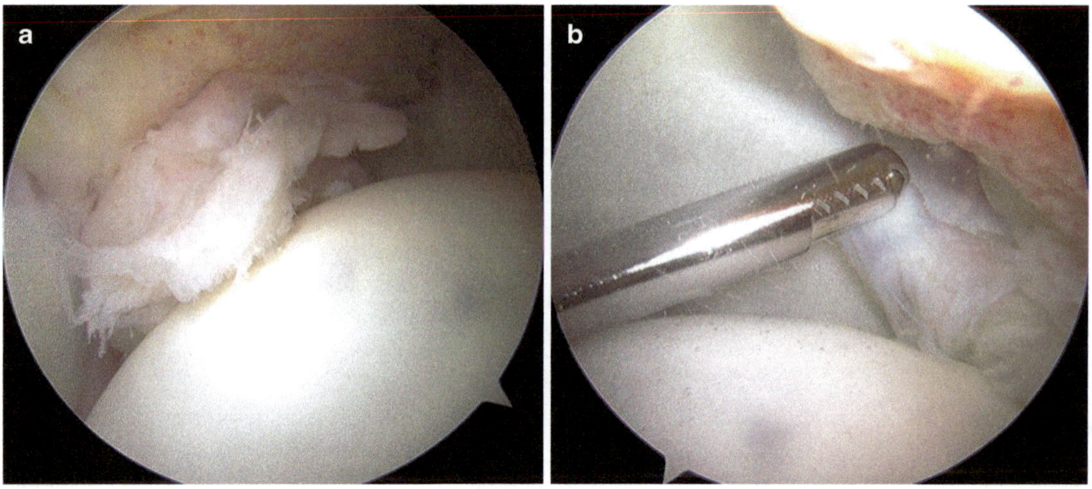

Fig. 7.7 (**a**) Arthroscopic view of a subtotal rupture of the ligamentum teres of the hip and (**b**) its progressive resection with an angulated shaver

drilled through the greater trochanter, exiting in the centre of the fovea. An acetabular tunnel is drilled through this previous femoral tunnel, with internal rotation and abduction of the hip in order to position the tunnel in the lower and slightly posteriorly portion of the acetabular fossa. A semitendinosus autograft is then pulled into the two tunnel with an endobutton positioned behind the acetabular tunnel. The graft is then fixed with an interference screw in the femoral tunnel under tension, with the hip fixed with 10° of hyperextension and 60° of external rotation.

Indications for LT reconstruction have yet to be clearly established. For now on, it seems reserved to patients with isolated complete LT tears presenting signs of micro-instability and increased external rotation [15].

7.4.3 Micro-Instability

Tears of the ligamentum teres can also occur in patients presenting micro-instability. This pathology is secondary to an excessive mobility of the hip, leading to chronicle repeated hip subluxations which will cause chondral and labral damages. This hypermobility results from an association of minor anomalies concerning the bony architecture and the periarticular soft tissues [20].

Concerning bony anomaly, it is mostly an insufficient acetabular anterior and/or lateral coverage angle between 25° and 20° called borderline dysplasia. Bony anomaly of the femur can be present with the association of coxa valga and augmented femoral anteversion. These bone architectural abnormalities induce higher mechanical stresses on the anterosuperior acetabular rim and labrum [21].

Concerning periarticular soft tissues, the capsule, the iliofemoral ligament and the ligamentum teres are the main structures fighting against anterior subluxation [22, 23]. In the cases of borderline dysplasia and hyperlaxity, ligamentum teres tears can occur next to an exaggerated or repeated hip extension and external rotation.

In the case of a complete ligamentum teres tear, even if ligamentum teres reconstruction can be proposed [15], we must keep in mind that the most important is to correct bony anomalies, in order to avoid recurrence. Thus, periacetabular osteotomy or shelf acetabuloplasty should be performed in case of acetabular dysplasia and femoral osteotomy in case of severe femoral abnormalities [21, 24, 25].

Arthroscopic capsular plication [26] (Fig. 7.8) or ligamentum teres reconstruction [27, 28] alone can be suggested only if bone anomalies

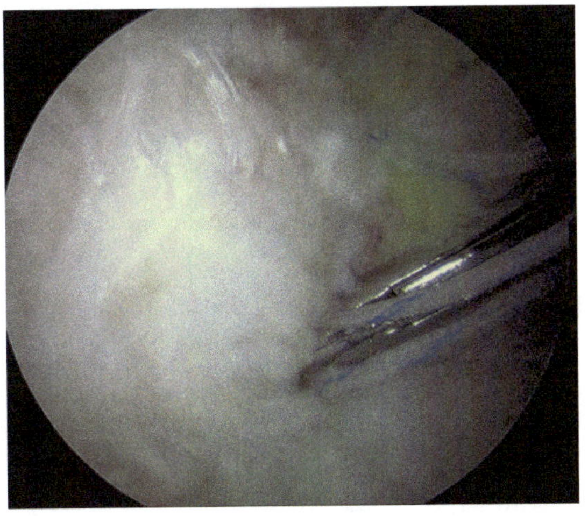

Fig. 7.8 Arthroscopic extra-articular view of a capsular plication with three endoscopic sutures

are minimal. The procedure will then focus on soft tissues in order to stiffen the hip. If preliminary results seem to be encouraging, these arthroscopic procedures are still to be evaluated with longer follow-up.

7.5 Coxarthrosis

Coxarthrosis is a chronicle hip disease of increasing frequency with age and anomalies in joint bone architecture. It is characterized by a thinning of the cartilage thickness induced by mechanical and inflammatory processes.

The patients mostly complain about pain and stiffness. Coxarthrosis pain is usually localized at the anterior face of the groin. It may radiate to the great trochanter, to the buttock, to the anterior side of the thigh or even to the knee. Knee pain with normal X-rays and a normal clinical examination is suggestive of hip origin. The patient can describe two different kinds of pain: mechanical or inflammatory pain.

There is no necessary close relationship between pain and X-ray anomalies. Patients can be very painful and a significant impact on daily activities with few X-ray signs, mostly in favour for an inflammatory process, while others will be painless with a full narrowing joint space. The decrease in cartilage thickness is qualified by the Tönnis classification from conventional hip X-rays: four grades are described, from grade 0 with no arthrosis sign to grade 3 with a full loss of joint space, osteophytes and increased deformity of the femoral head and acetabulum. MRI will allow diagnosing any joint or extra-articular inflammatory process.

To choose between a conservative treatment and a total hip replacement, we must take into account the age of the patient, the grade of osteoarthritis and the possible coxarthrosis risk factors (femoroacetabular impingement [29], hip dysplasia or synovial pathology):

- The oldest the patient is, the less effective the conservative treatment will be. This is induced by the fact that degenerative lesions are more common with advanced age. There is no true age limit for arthroscopy, but patients aged 50 years or older with 2 mm of joint space or less show early THR conversion [30, 31].
- A high grade of osteoarthritis, according to the Tönnis classification, at the time of the surgery is a bad prognosis factor for conservative hip surgery [32–34]. We consider Tönnis Grade 2 osteoarthritis as the limit for conservative treatment [35]. The radiological thickness of the joint space is also important to take into consideration. The best results for arthroscopic treatment are when the joint space thickness is more than 2 mm [36].
- Concerning risk factors for osteoarthritis induced by femoroacetabular impingement, the aim of the treatment is to remove all factors of impingement, osteophytes included, and to repair or reconstruct any labral tear because of its better prognosis [37, 38]. Cartilage lesions should be managed as discuss in Chap. 9.
- For osteoarthritis induced by dysplasia with lateral coverage angle inferior to 20°, the aim of conservative surgery is to restore a normal bone architecture. Surgery is realized in order to reduce stresses on the anterosuperior acetabular edge. Periacetabular osteotomy [39] or shelf acetabuloplasty [34, 40] should then be

considered. Some authors have proposed arthroscopic shelf acetabuloplasty [25].

- For inflammatory diseases, medical treatments and articular injections are possible according to rheumatologic instructions.

Conclusion

As seen in this chapter, several pathologies of the central compartment are providers of groin pain. The diagnosis of these pathologies are easier to establish nowadays, by the progress of medical imaging and improved biomechanical and medical knowledge.

If THR has great results concerning most of these pathologies in their advanced stage, conservative surgery and arthroscopic hip treatments are in constant development to better treat central compartment pathologies in their early stage and preserve bone capital.

References

1. Boyer T, Dorfmann H. Arthroscopy in primary synovial chondromatosis of the hip: description and outcome of treatment. J Bone Joint Surg Br. 2008;90(3):314–8.
2. Milgram JW. Synovial osteochondromatosis: a histopathological study of thirty cases. J Bone Joint Surg Am. 1977;59(6):792–801.
3. Lim SJ, Chung HW, Choi YL, Moon YW, Seo JG, Park YS. Operative treatment of primary synovial osteochondromatosis of the hip. J Bone Joint Surg Am. 2006;88(11):2456–64.
4. Marchie A, Panuncialman I, McCarthy JC. Efficacy of hip arthroscopy in the management of synovial chondromatosis. Am J Sports Med. 2011;39(Suppl):126S–31S.
5. Lequesne M, Mathieu P, Winkler P. La chondromatose synoviale de la hanche. Rev Rhum. 1999;66(Suppl):41–7.
6. Bouali H, Deppert EJ, Leventhal LJ, Reeves B, Pope T. Pigmented villonodular synovitis: a disease in evolution. J Rheumatol. 2004;31(8):1659–62.
7. Gonzalez Della Valle A, Piccaluga F, Potter HG, Salvati EA, Pusso R. Pigmented villonodular synovitis of the hip: 2- to 23-year followup study. Clin Orthop Relat Res. 2001;388:187–99.
8. Boyer T, Dorfmann H, Legré V, May O. Pathologie synoviale de la hanche. Rev Chir Orthop. 2010;96S:264–71.
9. May O, Boyer T, Dorfmann H. Pathologie du repli pectinéo-fovéal et traitement arthroscopique. Rev Chir Orthop. 2004;90(2):182.
10. El-Sayed AM. Treatment of early septic arthritis of the hip in children: comparison of results of open arthrotomy versus arthroscopic drainage. J Child Orthop. 2008;2(3):229–37.
11. Nusem I, Jabur MK, Playford EG. Arthroscopic treatment of septic arthritis of the hip. Arthroscopy. 2006;22(8):902–3.
12. Mullis BH, Dahners LE. Hip arthroscopy to remove loose bodies after traumatic dislocation. J Orthop Trauma. 2006;20(1):22–6.
13. Svoboda SJ, Williams DM, Murphy KP. Hip arthroscopy for osteochondral loose body removal after a posterior hip dislocation. Arthroscopy. 2003; 19(7):777–81.
14. Katz LD, Haims A, Medvecky M, McCallum J. Symptomatic hip plica: MR arthrographic and arthroscopic correlation. Skeletal Radiol. 2010;39(12):1255–8.
15. Lindner D, Sharp KG, Trenga AP, Stone J, Stake CE, Domb BG. Arthroscopic ligamentum teres reconstruction. Arthrosc Tech. 2013;2(1):e21–5.
16. O'Donnell J, Economopoulos K, Singh P, Bates D, Pritchard M. The ligamentum teres test: a novel and effective test in diagnosing tears of the ligamentum teres. Am J Sports Med. 2014;42(1):138–43.
17. Devitt BM, Philippon MJ, Goljan P, Peixoto LP, Briggs KK, Ho CP. Preoperative diagnosis of pathologic conditions of the ligamentum teres: is MRI a valuable imaging modality? Arthroscopy. 2014;30(5):568–74.
18. Datir A, Xing M, Kang J, Harkey P, Kakarala A, Carpenter WA, Terk MR. Diagnostic utility of MRI and MR arthrography for detection of ligamentum teres tears: a retrospective analysis of 187 patients with hip pain. AJR Am J Roentgenol. 2014;203(2):418–23.
19. Byrd JW, Jones KS. Traumatic rupture of the ligamentum teres as a source of hip pain. Arthroscopy. 2004;20(4):385–91.
20. Harris JD, Gerrie BJ, Lintner DM, Varner KE, McCulloch PC. Microinstability of the hip and the splits radiograph. Orthopedics. 2016;39(1): e169–75.
21. Klaue K, Durnin CW, Ganz R. The acetabular rim syndrome. A clinical presentation of dysplasia of the hip. J Bone Joint Surg Br. 1991;73(3):423–9.
22. Hewitt JD, Glisson RR, Guilak F, Vail TP. The mechanical properties of the human hip capsule ligaments. J Arthroplasty. 2002;17(1):82–9.
23. Myers CA, Register BC, Lertwanich P, Ejnisman L, Pennington WW, Giphart JE, LaPrade RF, Philippon MJ. Role of the acetabular labrum and the iliofemoral ligament in hip stability: an in vitro biplane fluoroscopy study. Am J Sports Med. 2011;39(Suppl):85S–91S.
24. Sankar WN, Millis MB, Kim YJ. Instability of the hip in patients with Down Syndrome: improved results with complete redirectional acetabular osteotomy. J Bone Joint Surg Am. 2011;93(20):1924–33.

25. Uchida S, Wada T, Sakoda S, Ariumi A, Sakai A, Iida H, Nakamura T. Endoscopic shelf acetabuloplasty combined with labral repair, cam osteochondroplasty, and capsular plication for treating developmental hip dysplasia. Arthrosc Tech. 2014;3(1):e185–91.
26. Domb BG, Stake CE, Lindner D, El-Bitar Y, Jackson TJ. Arthroscopic capsular plication and labral preservation in borderline hip dysplasia: two-year clinical outcomes of a surgical approach to a challenging problem. Am J Sports Med. 2013;41(11):2591–8.
27. Cerezal L, Arnaiz J, Canga A, Piedra T, Altonaga JR, Munafo R, Perez-Carro L. Emerging topics on the hip: ligamentum teres and hip microinstability. Eur J Radiol. 2012;81(12):3745–54.
28. Domb BG, Philippon MJ, Giordano BD. Arthroscopic capsulotomy, capsular repair, and capsular plication of the hip: relation to atraumatic instability. Arthroscopy. 2013;29(1):162–73.
29. Ganz R, Parvizi J, Beck M, Leunig M, Notzli H, Siebenrock KA. Femoroacetabular impingement: a cause for osteoarthritis of the hip. Clin Orthop Relat Res. 2003;417:112–20.
30. Domb BG, Linder D, Finley Z, Botser IB, Chen A, Williamson J, Gupta A. Outcomes of hip arthroscopy in patients aged 50 years or older compared with a matched-pair control of patients aged 30 years or younger. Arthroscopy. 2015;31(2):231–8.
31. Philippon MJ, Schroder ESBG, Briggs KK. Hip arthroscopy for femoroacetabular impingement in patients aged 50 years or older. Arthroscopy. 2012;28(1):59–65.
32. Daivajna S, Bajwa A, Villar R. Outcome of arthroscopy in patients with advanced osteoarthritis of the hip. PLoS One. 2015;10(1):e0113970.
33. Gicquel T, Gedouin JE, Krantz N, May O, Gicquel P, Bonin N. Function and osteoarthritis progression after arthroscopic treatment of femoro-acetabular impingement: a prospective study after a mean follow-up of 4.6 (4.2–5.5) years. Orthop Traumatol Surg Res. 2014;100(6):651–6.
34. Rosset P, Heudel B, Laulan J, Garaud P, Favard L. Long-term evolution following shelf procedure for hip dysplasia in adults. Shelf survival analysis in 68 cases and retrospective review of 44 with at least 26 years follow-up. Acta Orthop Belg. 1999; 65(3):315–26.
35. Domb BG, Gui C, Lodhia P. How much arthritis is too much for hip arthroscopy: a systematic review. Arthroscopy. 2015;31(3):520–9.
36. Skendzel JG, Philippon MJ, Briggs KK, Goljan P. The effect of joint space on midterm outcomes after arthroscopic hip surgery for femoroacetabular impingement. Am J Sports Med. 2014; 42(5):1127–33.
37. Domb BG, El Bitar YF, Stake CE, Trenga AP, Jackson TJ, Lindner D. Arthroscopic labral reconstruction is superior to segmental resection for irreparable labral tears in the hip: a matched-pair controlled study with minimum 2-year follow-up. Am J Sports Med. 2014;42(1):122–30.
38. Larson CM, Giveans MR, Stone RM. Arthroscopic debridement versus refixation of the acetabular labrum associated with femoroacetabular impingement: mean 3.5-year follow-up. Am J Sports Med. 2012;40(5):1015–21.
39. Chen M, Shang XF. Surgical treatment for young adult hip dysplasia: joint-preserving options. Int Orthop. 2015;40(5):891–900.
40. Hirose S, Otsuka H, Morishima T, Sato K. Long-term outcomes of shelf acetabuloplasty for developmental dysplasia of the hip in adults: a minimum 20-year follow-up study. J Orthop Sci. 2011;16(6):698–703.

8 Hip Cartilage Lesions

Piero Volpi, Raul Zini, Alessandro Quaglia, and Manlio Panascì

8.1 Introduction

It is well known that the articular cartilage has poor reparative capacities, and this is mainly due to the absence of vascularization, which would allow the progenitor cells to colonize the tissue and then repair the injury [1, 2].

The dependent mechanical stress of the cartilage itself causes damage to chondrocytes and extracellular matrix by triggering a process of degeneration limited at first and then involving the adjacent cartilage through an increased stress on the perilesional healthy tissue [3].

If within certain dimensions in fact the surrounding healthy tissue is able to absorb the loads of the damaged area, over a certain size, this can no longer take place by triggering a load of the lesion evolutionary process [4].

It becomes very important to treat focal cartilage lesions in order to resolve the pain, to regain joint function, to return to normal activities, and to limit the arthritic degeneration of the joint.

The articular cartilage of the hip is generally thinner than the cartilage of the knee with an average thickness which can oscillate, on average, between 1.08 and 2.4 mm at the level of the femoral head and between 1.24 and 2.25 mm to the acetabulum [5].

The thickness reduction could be explained by the better hip joint congruity and the sealing effect of the acetabular labrum, resulting in a smaller shock absorbing request [6]. The better "performance" with respect to the knee joint may partly explain the fact that chondral lesions in the absence of concomitant pathologies rarely occur [7].

8.2 Pathogenesis

Cartilage lesions are often associated with other diseases, primarily femoroacetabular impingement (FAI) and labral tears.

Therefore, hip focal chondral lesions could be due to trauma. Typically, impact mechanisms on the great trochanter side can generate cartilage lesions at the level of the femoral head and at the level of the acetabulum. The reduced ability to absorb energy due to its subcutaneous position explains its biomechanical mechanism [8].

FAI could be a cam-type impingement or a pincer type. Recent studies have also shown a

P. Volpi • A. Quaglia
Ortopedia del Ginocchio e Traumatologia dello Sport, Humanitas Research Hospital, Rozzano, Milan, Italy

R. Zini
Maria Cecilia Hospital GVM Care and Research, Cotignola, Ravenna, Italy

M. Panascì (✉)
San Carlo di Nancy Hospital - GVM Care and Research, Rome, Italy
e-mail: manliopa@gmail.com

R. Zini et al. (eds.), *Groin Pain Syndrome*, DOI 10.1007/978-3-319-41624-3_8

higher genetic expression of pro-inflammatory cytokines at the level of the articular cartilage in patients diagnosed with FAI compared to patients with osteoarthritis as well confirming the theory that the FAI can be a trigger for an early joint degeneration [9].

In cam-type impingement, the altered form of the head no longer spherical creates a conflict with the acetabulum, in particular in the phases of flexion and intra-rotation of the hip, creating the acetabular cartilage lesions and subsequent possible lesions of the labrum [10].

PINCER type is a localized (acetabular retroversion) or generalized (coxa profunda and acetabuli protrusio) acetabular overcoverage. The increased contact surface between femoral epiphysis and acetabulum determines an early contact between acetabular rim and femoral neck, thus causing an early degeneration and tear of the acetabular labrum.

According to some authors, also hip chondral lesions are often associated with lesions of the acetabular labrum. The most affected locations are the area above the acetabulum (24%) and the rear area (25%) [11]. The earliest chondral lesion detectable at hip arthroscopy is a chondral split, progressing onward to a more formal delamination, chondral flap lesion, and subchondral cyst. Other conditions that can often be associated with hip chondral lesions are the presence of loose bodies, hip subluxations, dysplastic hip, or osteonecrosis of the femoral head [7, 12, 13].

8.3 Clinic

Symptoms that may occur in the hip chondral lesions are often nonspecific.

Patients report, as a main symptom, groin pain or buttock pain, sometimes in other locations with irradiation. They may also present joint stiffness and snaps commonly in the presence of free articular fragments or mobile bodies [14].

Usual sports activities may be limited, and sometimes even simply sitting position or hip flexion exercises can, in the presence of FAI, exacerbate pain [14].

The timing of the onset of symptoms also can help to understand if the patient has post-traumatic or degenerative pathology.

The clinic is often quite poor in elements; pain, especially if associated with a FAI condition, is evoked in flexion tests, adduction, and internal rotation (FADIR test) or in flexion, abduction, and external rotation (FABER test).

8.4 Classification

The most used classification in hip chondral lesions are mainly four:

1. Outerbridge classification, originally described for cartilage lesions of the knee, is divided into four degrees (Table 8.1) [15].
2. ICRS classification, similar to the first, is divided into five degrees that evolve from grade 0 to grade 4 evaluating, above all, the depth of the lesion [16].
3. Back classification is divided into six degrees specific for chondral lesions of the hip [17].
4. Sampson classification is divided into two groups (acetabular and femoral), with a focus on chondral delamination in the acetabular side of the hip (Table 8.2) [18].

Table 8.1 The Outerbridge classification

Grade	Size	Description
I		Softening/swelling
II	≤1.3 cm (1/2 in.)	Fragmentation/fissuring
III	>1.3 cm (1/2 in.)	
IV		Erosion/subchondral bone exposure

Table 8.2 The Sampson classification

Acetabular damage classification
AC 0–5 no damage
AC 1 softening with no wave sign
AC 1w softening with wave sign and intact labro-cartilage junction
AC 1wTj softening with wave sign and torn labro-cartilage junction
AC 1wD softening with wave sign and intact labro-cartilage junction with delamination
AC 1wTjD softening with wave sign and torn labro-cartilage junction with delamination
AC 2 fibrillation
AC 2Tj fibrillation with torn labro-cartilage junction
AC 3 exposed bone small area <1 cm^2
AC 4 exposed bone large area >1 cm^2
Femoral damage classification
Intact head substrate bone with chondral damage (no avascular necrosis)
HC 0 no damage
HC 0T uniform thinning (T)
HC 1 softening
HC 2 fibrillation
HC 3 exposed bone
HC 4 any delamination
Special class: HTD traumatic defect (size in mm)
HDZ demarcation zone from FAI

Abbreviations: *A* acetabulum, *C* cartilage defects, *D* with delamination, *Tj* Torn labro-cartilage junction, *w* with wave sign, *HC* femoral head cartilage, *T* thinning, *TD* traumatic defect, and *DZ* demarcation zone from FAI

8.5 Imaging

All young patients with hip pain should be evaluated by basic standards such as AP X-ray. In FAI pincer type, because of an acetabular retroversion, it could be observed the crossover sign is due to the projection of the anterior acetabular wall that intersects with the projection of the rear acetabular wall [19].

Axial projection of the hip is also very important to evaluate the possible presence of FAI cam. In particular, the most used axial projections are axial cross-table [20], the frog view or cross-leg projection, and the projection of Dunn at 45° [21].

Performing an arthro-RM, by infiltration of the contrast, and using a high-quality image (1.5 or 3 T) can show directly the chondral injury (Fig. 8.1) and possible associations of labral tears. The MRI can also show bone marrow edema or cystic modification, and this allows detection of overloaded areas or a subchondral bone expression.

The presence of edema under the cartilage injury can also be due to a recent trauma.

The delayed gadolinium-enhanced MRI cartilage (dGEMRIC) is often used to assess the biochemical structure and thus the quality of articular cartilage, but this is rarely possible in clinical practice.

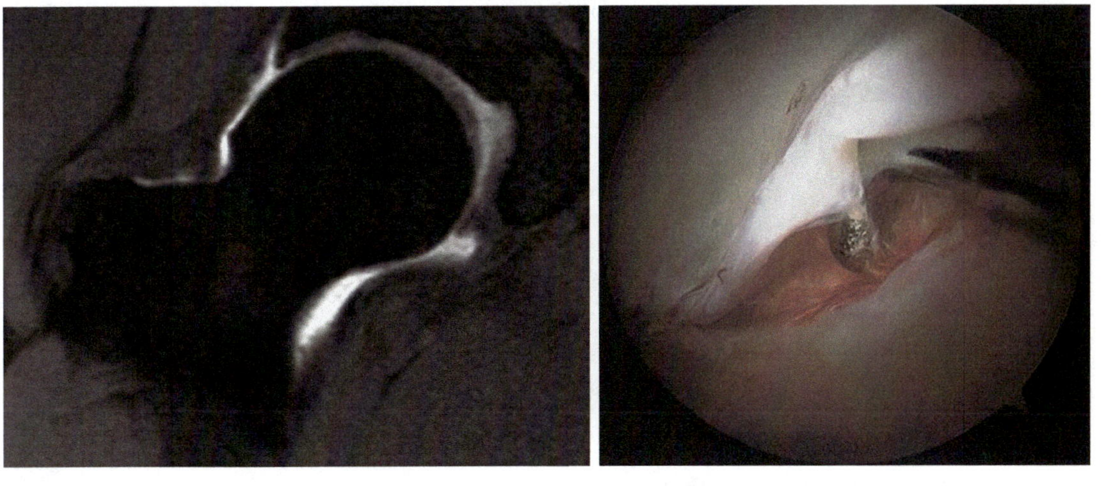

Fig. 8.1 *Left*: Acetabular cartilage delamination ("inverse Oreo sign") on arthro-MRI. *Right*: arthroscopic view showing interruption of the chondrolabral junction

It is possible, for diagnosis, to perform intra-articular infiltration of anesthetic.

8.6 Treatment

Debridement is the most common treatment for chondral defects, especially in the low-grade arthritis. Since the severity of the lesion affects significantly the outcome of the surgery, it is necessary to act as early as possible [22].

Simple debridement of the joint surfaces is carried out through the use of a shaver, to provide a smooth shearing surface for the movement of the joint. Results have demonstrated that 30–60% of patients undergoing debridement reported good postoperative clinical outcomes and the conversion to a THA is significantly delayed [23].

When the labro-cartilage junction is intact, several ways to treat it are possible. Injecting fibrin glue beneath the defect may treat the lesion, by causing adherence. Some studies report that the defect is helped from microfractures to obtain a healing response. Elevating the full-thickness defect, curetting and microfracturing the acetabular subchondral bone may stimulate the biologic healing response beneath the full-thickness cartilage [24].

The microfractures consist in digging with an awl some holes, perpendicular to the cartilage surface, with a diameter of about 4 mm, leaving this same much space between one another. The goal of microfracture is to bring marrow cells and growth factors from the underlying bone marrow into the affected chondral defect. By penetrating the subchondral bone, pluripotent marrow cells can develop and form new fibrocartilage to fill the chondral defect (Fig. 8.2). It is reported in the literature that microfracture provides healing of small lesions (between 1.5 and 3 cm^2) in more than 90% of cases (7) (8) with a microscopically good-quality cartilage. Conversely, scarce effectiveness in Tönnis grade 3 or Outerbridge 3–4 lesions has been demonstrated [25].

With labro-cartilage junction tear, we routinely excise the unstable portion of the articular cartilage flap and stimulate subchondral bone through nano-microfracture system, a subchondral needling procedure which disrupts significantly less surface area than a standard

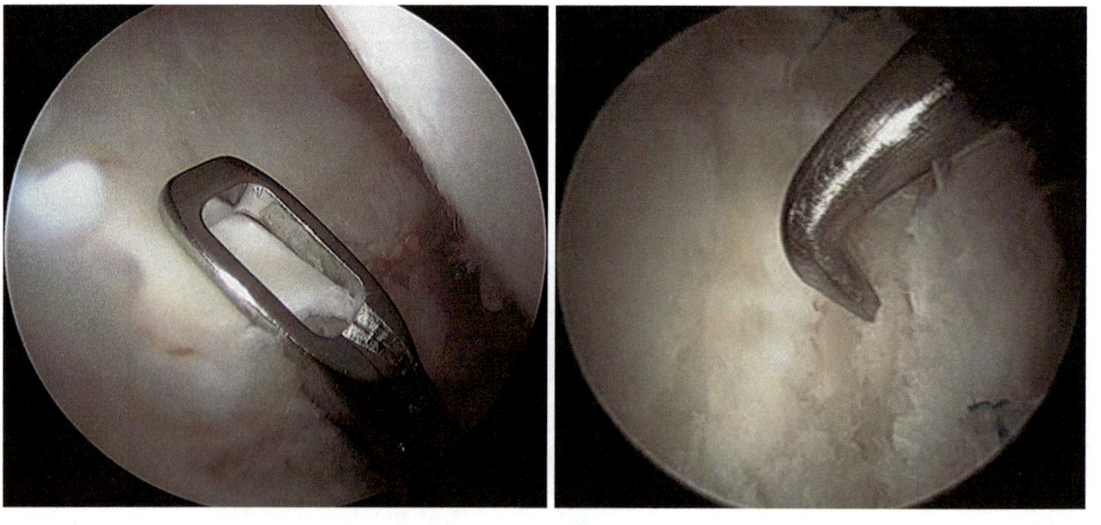

Fig. 8.2 *Left*: Cartilage is removed up to a stable margin. *Right*: microfractures are performed after repairing the labrum (the suture is visible on the bottom right)

microfracture technique to reduce damage to the subchondral bone plate (Fig. 8.3).

One other effective treatment steady nowadays is the autologous chondrocyte implantation (ACI). This technique can be carried out only at some conditions: the lesion must be a full-thickness tear and the subchondral bone must be intact. Lesions larger than 3 cm can be treated, but usually smaller than 10 cm in width. The ACI-MACI technique is performed only through the help of a biology lab, because of the need of a culture-based growth of the chondrocytes harvested from a joint (other than the involved hip) of the patient. Once the chondrocyte cultivation has growth enough to fill the lesion, it is implanted. Last updates in this technique involve the use of a scaffold ("matrix," from this MACI, "matrix-assisted ACI") to set the correct delivery of the cells on the lesion (Fig. 8.4). This technique can be performed arthroscopically, in contrast with the simple ACI (in which a patch and an injectable solution are used). Although this technique may be considered a shared option for the treatment of mild to severe lesions, it has not been demonstrated significantly better than simple debridement neither for early nor later grades of OA. Management of the cartilage delamination can involve both these techniques, once the loose portion of the defect has been removed. However, it has been reported that also direct suture or fibrin adhesion could be provided for small lesions if the delaminated cartilage appears healthy [26].

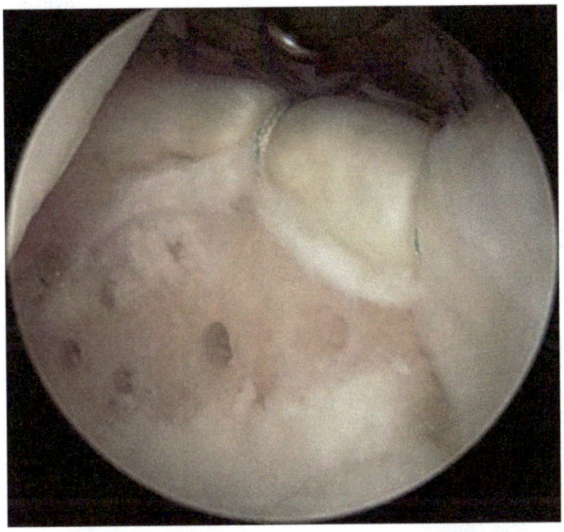

Fig. 8.3 Nano-fracture technique after labral refixation

Damage only on the femoral side is mainly due to traumatic events. The damage may be caused from impaction of the head to the notch and/or damage seen from dislocation.

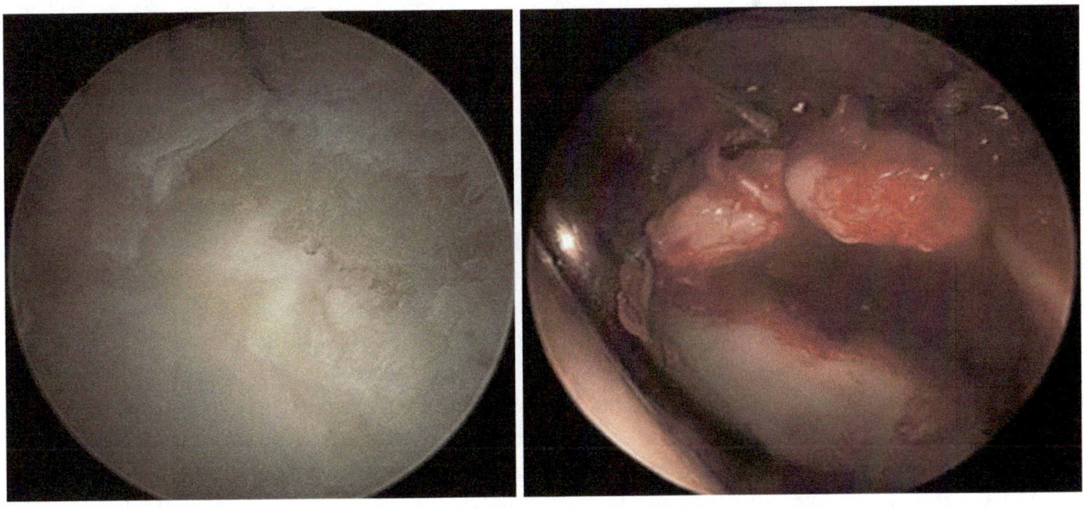

Fig. 8.4 Chondral damage before (*left*) and after a chitosan-glycerol phosphate/blood implant

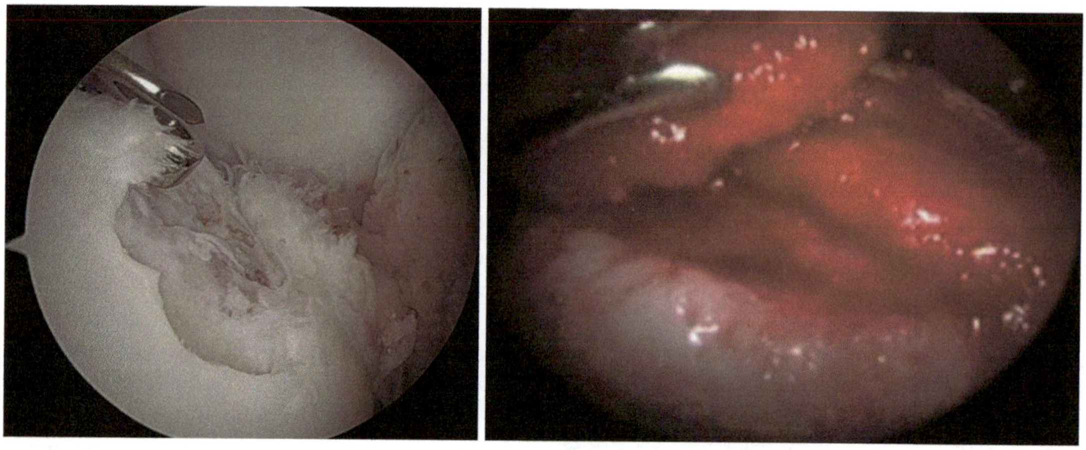

Fig. 8.5 *Left*: Chondral defect of a femoral head. *Right*: ACI technique to fill the defect

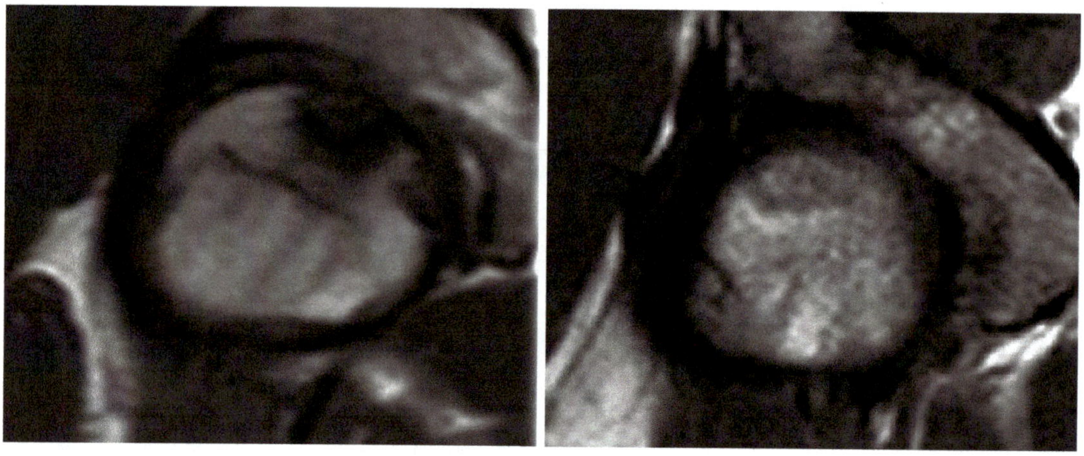

Fig. 8.6 *Left*: MRI showing the femoral head defect. *Right*: The defect is no more visible 18 months after the arthroscopic procedure

Arthroscopic excision of the fragment may be sufficient; however, in case of a localized damage on weight bearing area, we usually fill the gap with curettage and microfracture or chondral plugs (Figs. 8.5 and 8.6).

8.7 Outcome

Studies on outcomes after arthroscopic correction of FAI have found negative influences of OA on postoperative outcomes [26, 27]. Preoperative radiographic joint space narrowing, together with preoperative MHHS score and labral debridement or repair, predicted 35% of postoperative MHHS, and generalized severe cartilage lesions intraoperatively led to high rates of arthroplasty within 3 years of arthroscopy [28]. Increasing preoperative joint space narrowing has been shown to be an independent predictor for poorer outcome with respect to postoperative MHHS scores and undergoing arthroplasty in a recent cohort undergoing arthroscopy for FAI. Despite degree of OA being inversely related to amount of improvement, some data show instead that patients with moderate to severe articular cartilage degeneration or rim lesions improve outcome scores 12 months after surgery, but some results provide preliminary support for the benefits of hip arthroscopy even in patients with more severe OA. Other studies have shown that in patients with already marked generalized chondral lesions, arthroscopy does not have any effect beyond the short-term pain relief resulting from debridement, underlining the fact that FAI with advanced osteoarthrosis, particularly Tönnis grade 3, is not an indication for arthroscopic FAI correction [29, 30].

In two studies, Philippon et al. report on the outcome of the arthroscopic treatment of FAI and reveal that patients improve in terms of PROMs, but that they had an increased rate of conversion to THA with any joint space of 2 mm or less. They reported a THA conversion rate of 9 and 20%, respectively [30]. In 2014, Gicquel et al. reported on 53 hips treated arthroscopically for FAI [31]. Patients with Tönnis grade 1 hips had lower satisfaction rates and a higher conversion rate to THA than patients with Tönnis grade 0. In a systematic review by Kemp et al., it is concluded that patients with hip OA report positive outcomes from hip arthroscopy. Patients with hip OA had inferior outcomes compared with those with no hip OA. The severity of chondropathy and higher patient age were associated with a higher risk and more rapid progression to THA.

The treatment of chondral lesions of the hip is difficult and may not ever reestablish normal chondral anatomy; however, the methods described will possibly improve pain and function of damaged hips. This new method of treating full-thickness acetabular articular cartilage defects offer a good solution for cartilage repair while attempting to preserve most of the cartilage matrix in the process. The value of arthroscopic techniques for primary symptoms and/or prevention of OA remains unclear. Randomized controlled trials to evaluate efficacy are needed especially to understand the physiological response and imaging characteristics after microfracture or plug positioning.

References

1. Vinatier C, Mrugala D, Jorgensen C, et al. Cartilage engineering: a crucial combination of cells, biomaterials and biofactors. Trends Biotechnol. 2009;27:3017–314.
2. de Girolamo L, Bertolini G, Cervellin M, et al. Treatment of chondral defects of the knee with one step matrix-assisted technique enhanced by autologous concentrated bone marrow: in vitro characterization of mesenchymal stem cells from iliac crest and subchondral bone. Injury. 2010;4:1172–7.
3. Colwell CW, D'Lima DD, Hiennecke HR, et al. In vivo changes after mechanical injury. Clin Orthop Relat Res. 2001;391S:S116–23.
4. Guettler JH, Demetropoulos CK, Yang KH, et al. Osteochondral defects in the Human knee: influence of defect on cartilage rim stress and load redistribution to surrounding cartilage. Am J Sports Med. 2004;32:1451–8.
5. Shepherd DET, Seedhom BB. Thickness of human articular cartilage in joints of the lower limb. Ann Rheum Dis. 1999;58:27–34.
6. Simon WH, Friedenberg S, Richardson S. Joint congruence. A correlation of joint congruence and thickness of articular cartilage in dogs. J Bone Joint Surg Am. 1973;55-A:1614–20.
7. Philippon MJ. The hip, cartilage E-book. In: Kon E, Berruto M, Condello V, Peretti GM, Ronga M, editors. Ed CIC. p. 416–26.
8. Byrd JWT. Lateral impact injury. A source of occult hip pathology. Clin Sports Med. 2001;20:801–15.
9. Chinzei N, Hashimoto S, Fujishiro T, et al. Inflammation and degeneration in cartilage samples for patient with femoroacetabular impingement. J Bone Joint Surg Am. 2016;98(2):135–41.
10. Beck M, Kalhor M, Leuning M, Ganz R. Hip morphology influences the pattern of damage to the acetabular cartilage: femoroacetabular impingement as cause of early osteoarthritis of the hip. J Bone Joint Surg Br. 2005;87(7):1012–8.
11. McCarthy JC, Lee JA. Arthroscopic intervention in early hip disease. Clin Orthop Relat Res. 2004;429:157–62.
12. Deangelis NA, Busconi BD. Assessment and differential diagnosis of painful hip. Clin Orthop Relat Res. 2003;406:11–8.
13. Schenker ML, Martin R, Weiland DE, Philippon MJ. Current trends in hip arthroscopy: a review of injury diagnosis, techniques and outcome scoring. Curr Opin Othop. 2005;16:89–94.
14. Beaulè PE, O'Neil M, Rakhra K. Acetabular labral tears. J Bone Joint Surg Am. 2003;91(3):701–10.
15. Outerbridge RE. The etiology of chondromalacia patellae. J Bone Joint Surg Br. 1961;43-B:752–7.
16. ICRS cartilage Injury Evaluation package. 2000. http://www.cartialge.org/Evaluation Package/ICRS Evaluation.pdf
17. Back M, Leuning M, Parvizi J, et al. Anterior femoroacetabular impingement. Part II. Midterm results of surgical treatment. Clin Othop Relat Res. 2004;418:67–73.
18. Sampson TG. Arthroscopic treatment for chondral lesions of the hip. Clin Sports Med. 2011;30:331–48.
19. Siebenrock KA, Schoeniger R, Ganz R. Anterior femoroacetabular impingement due to acetabular retroversion. Treatment with periacetabular osteotomy. J Bone Joint Surg Am. 2003;85-A(2):278–86.
20. Eijer H, Leuning M, Mahomed MN, et al. Cross-table lateral radiograph for screening of anterior femoral head-neck offset in patients with femoroacetabular impingement. Hip Int. 2001;11:37–41.
21. Meyer DC, Beck M, Ellis T, et al. Comparison of six radiographic projections to assess femoral head asphericity. Clin Orthop Relat Res. 2006;445:181–5.

22. McCarthy JC, Noble PC, Schuck MR, Wright J, Lee J. The watershed labral lesion: its relationship to early arthritis of the hip. J Arthroplasty. 2001;16(8 Suppl 1):81–7.
23. Espinosa N, Rothenfluh DA, Beck M, Ganz R, Leunig M. Treatment of femoroacetabular impingement: preliminary results of labral refixation. J Bone Joint Surg Am. 2006;88(5):925–35.
24. El Bitar YF, Lindner D, Jackson TJ, Domb BG. Joint-preserving surgical options for management of chondral injuries of the hip. J Am Acad Orthop Surg. 2014;22(1):46–56. PubMed PMID: 24382879
25. Karthikeyan S, Roberts S, Griffin D. Microfracture for acetabular chondral defects in patients with femoroacetabular impingement: results at second-look arthroscopic surgery. Am J Sports Med. 2012;40(12):2725–30. PubMed PMID: 23136178
26. Philippon MJ, Schenker ML, Briggs KK, Maxwell RB. Can microfracture produce repair tissue in acetabular chondral defects? Arthroscopy. 2008;24(1):46–50.
27. Sekiya JK, Martin RL, Lesniak BP. Arthroscopic repair of delaminated acetabular articular cartilage in femoroacetabular impingement. Orthopedics. 2009; 32(9):692–6.
28. Stafford GH, Bunn JR, Villar RN. Arthroscopic repair of delaminated acetabular articular cartilage using fibrin adhesive. Results at one to three years. Hip Int. 2011;21(6):744–50. PubMed PMID: 22117261
29. Tzaveas AP, Villar RN. Arthroscopic repair of acetabular chondral delamination with fibrin adhesive. Hip Int. 2010;20(1):115–9. PubMed PMID: 20235074
30. Philippon MJ, Briggs KK, Yen YM, Kuppersmith DA. Outcomes following hip arthroscopy for femoroacetabular impingement with associated chondrolabral dysfunction: minimum two-year follow-up. J Bone Joint Surg Br. 2009;91:16–23.
31. Gicquel T, Gédouin JE, Krantz N, May O, Gicquel P, Bonin N; Function and osteoarthritis progression after arthroscopic treatment of femoro-acetabular impingement: a prospective study after a mean follow-up of 4.6 (4.2-5.5) years. Orthop Traumatol Surg Res. 2014;100(6):651–6.

9 Extra-articular Hip Pathology

Raul Zini and Manlio Panascì

9.1 Introduction

The snapping hip, according to the region concerned, is classified in three categories: external, internal, and intra-articular. The conflict of iliotibial band (ITB) with the greater trochanter region determines the hip external snap. Intra-articular pathologies in differential diagnosis with internal snapping hip are due to articular loose bodies, labral injury, or articular instability [1, 2]. The underlying causes of a conflict or mechanisms that may determine the onset of a snapping hip are shown in Table 9.1.

9.2 External Snapping Hip

The most common cause of snapping hip syndrome is irritation of the greater trochanter by the ITB. The ITB is a large flat tendinous structure that originates on the anterior superior portion of the iliac crest, crosses over the greater trochanter of the femur, and inserts onto the lateral condyle of the tibia.

When the hip is extended, the ITB is posterior to the greater trochanter. As the hip moves into flexion, the ITB moves anterior to the greater trochanter. Ordinarily, it glides smoothly over the greater trochanter with assistance from the underlying bursae [3]. When the posterior portion of the iliotibial tract or the anterior border of the gluteus maximus becomes thickened, however, this results in snapping of the tendon over the greater trochanter. The bursae can then become inflamed and further exacerbate the condition [4]. Coxa vara may predispose to a snapping hip. Other predisposing factors are hyperplasia of the trochanteric bursa, narrower bi-iliac width, prominent greater trochanters, and increased distance between the greater trochanters. A case of snapping hip secondary to

Table 9.1 Snapping hip

Internal snapping hip
Tendon impingement at the level of iliopectineal eminence
Iliopsoas impingement with the acetabular component in a THA
Tendon snap at the level of the upper branch of the pubic bone
Conflict with the anterior acetabular margin
Conflict between two components of a bifid tendon
Impingement of the tendon at the level of the anterior inferior iliac spine
External snapping hip
Thickening of the posterior ITB or the gluteus maximus
Trochanter deformities
Big offsets in THA
Intra-articular snapping hip
Osteochondromatosis, fragment results of articular fractures, loose bodies
Labral tears
Instability

R. Zini
Maria Cecilia Hospital GVM Care and Research, Cotignola, Ravenna, Italy

M. Panascì, M.D. (✉)
San Carlo di Nancy Hospital - GVM Care and Research, Rome, Italy
e-mail: manliopa@gmail.com

R. Zini et al. (eds.), *Groin Pain Syndrome*, DOI 10.1007/978-3-319-41624-3_9

fibrosis of the band/muscle related to multiple intramuscular injections has also been reported [5].

9.2.1 Physical Examination

External snapping hip is seen in athletes who undergo repetitive knee flexion, such as runners, dancers, and cyclists. Athletes will have pain over the greater trochanter of the femur, the lateral thigh, or radiating pain down to the knee. Patients often report hip instability symptoms. If severe enough, the snapping sensation will occur during normal ambulation. Once this area becomes inflamed, running or rising from a seated position may hurt continuously. On physical examination, in addition to evidence of the shot, which is often caused voluntarily by the patient, it can consist of tenderness upon insertion of IBD or the greater trochanter, with occasional painful side irradiation in the thigh. Patients may demonstrate a positive Ober test because of increased tension of the ITB.

9.2.2 Imaging

Diagnosis of snapping hip is a clinical one. Although dynamic ultrasound best demonstrates the snapping hip, cross-sectional imaging with MRI can demonstrate findings associated with external snapping hip, namely, the thickened iliotibial band or the muscle, and the associated secondary change of atrophy of the rest of the gluteus maximus muscle. Both signs should be looked for to confirm the diagnosis. Any intra-articular pathologies should be investigated with plain X-rays and MRI which could also reveal a trochanteric bursitis or a tear of the gluteus medius or minimus.

9.2.3 Treatment

This condition can be asymptomatic and however in some patients can cause pain and disability. Conservative therapy is, in most cases, decisive; physical therapy, with emphasis on stretching, strengthening, and alignment, can often help. Sometimes, treatment with a corticosteroid injection to the area can relieve inflammation. However, when the remission of symptoms is unsatisfactory, a surgical approach is indicated. A pioneer in this field was James Glick, who described the lateral approach and was also key to the development of specialized instrumentation required for this procedure. In 2006 Ilizaliturri was the first to underline comparable results between endoscopic and open procedures (Picture 9.1). We have recently published a modified endoscopic iliotibial band release with excellent results in terms of snapping phenomenon resolution, patient satisfaction, and return to previous level of activity [6, 7].

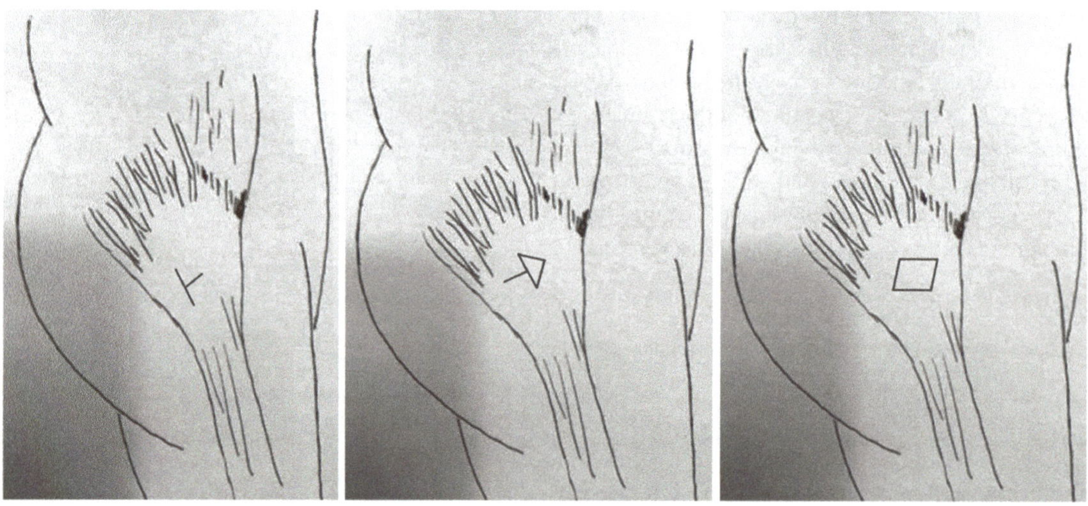

Picture 9.1 The Ilizaliturri technique

9.2.4 Surgical Technique

Both lateral and supine position can be used for the treatment of the external snapping hip. We usually prefer the supine position when a concomitant hip arthroscopy is performed (Picture 9.2). When using the lateral decubitus position, prepping and draping of the operating area and the leg is carried out in the standard fashion. The leg is draped so that the hip can be moved during surgery. With the operating area ready, the greater trochanter (GT) is outlined, and a spine needle is used to locate and to mark the apex of GT. Two portals, one anterior and one posterior, are created at the level of the snapping tract of the ITB. The 30° 4-mm scope is then introduced through the posterior portal, and water inflow with an arthroscopy pump at low pressure is started to develop a space between the subcutaneous tissue and the ITB. The space superficial to the ITB is further developed with the shaver until the ITB can be easily identified. A hooked radiofrequency (RF) probe is introduced trough the anterior portal, and hemostasis is carried out on the subcutaneous tissue if necessary. A horizontal cut is made into the ITB and a defect is created into the ITB. The trochanteric bursa, if inflamed, can easily be resected through the defect using a shaver and a RF ablator (Picture 9.3). The snapping should be tested at different times during the procedure so that adequate resection is obtained. It is appropriate at the end of the

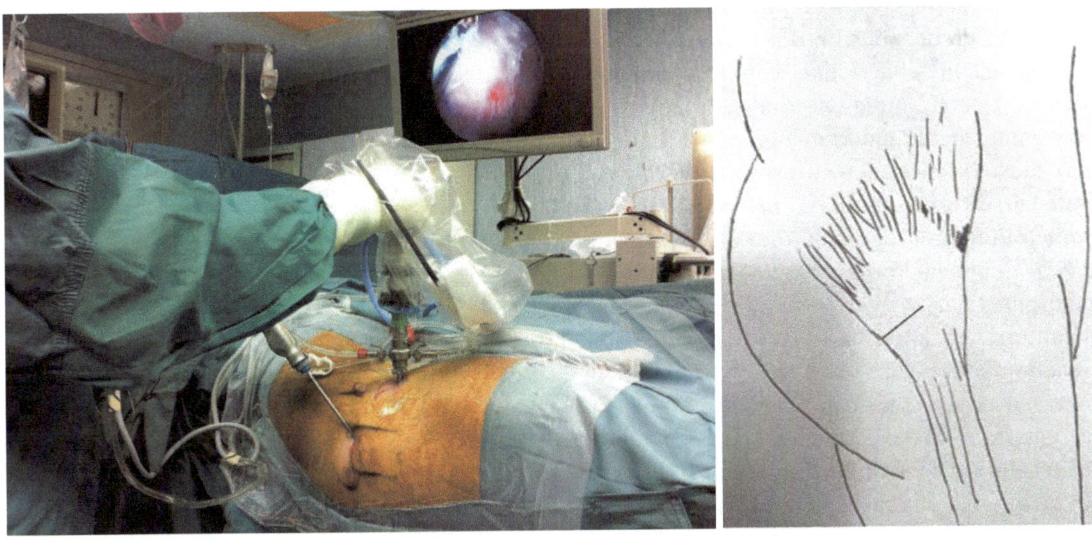

Picture 9.2 *Left*: the supine position can be easily used to perform the ITB release. *Right*: personal technique with horizontal section of the fibers

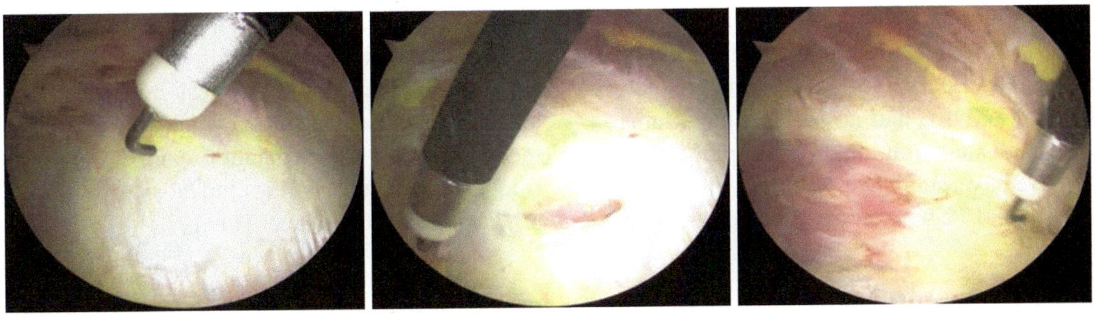

Picture 9.3 From *left* to *right*: an horizontal cut is made and a defect is created into the ITB

procedure to perform a coagulation of small vessels and cut and place a drain for about 12 h [8, 9].

9.2.5 Outcome

Physicians agree that, in the majority of patients, a conservative approach will be successful; only a select few patients will require operative intervention. The endoscopic approach allows to reach the same or better results in terms of snapping and pain resolution with lower incidence of local complication. The most frequently reported complication in the literature has been the incomplete relief of symptoms after operative intervention. This can be minimized with an accurate preoperative diagnosis; it also is fundamental to test the snap with provocative maneuvers at different times during the procedure so that an adequate resection could be obtained with particular attention to detect any focal thickening at the anterior edge of the gluteus maximus. In case of pain association, a complete bursectomy should be performed. There is limited literature regarding the results of endoscopic treatment for the external snapping hip syndrome, but early reports are encouraging. Ilizaliturri et al. created a diamond-shaped resection and release; after which, the trochanteric bursa was debrided. Ten of 11 hips (10 patients) were relieved of their snapping, and all were relieved of pain at a minimum of 1-year follow-up. No patient required revision surgery, and all patients returned to preoperative levels of activity [6]. Polesello et al. performed a gluteus maximus tenotomy to decrease tension on the iliotibial band. Seven of 9 hips (8 patients) were relieved of pain and snapping postoperatively, with 1 patient requiring a revision operation. All patients returned to their preoperative level of activity, and no patients complained of weakness at a minimum follow-up of 22 months [9]. We have published a study with 15 patients (3 men and 12 women) with symptomatic external snapping hip treated with an endoscopic release of the iliotibial band. The average age was 25 years (range 16–37 years). VAS score was significantly reduced with respect to the preoperative value with 60% of the patients pain-free. No revision procedures were indicated and all the patients returned to their previous level of activity [10].

9.3 Internal Snapping Hip

The treatment of the different pathologies of the iliopsoas tendon, in the case of relapsing tendonitis for conflict with adjacent structures or arising from problems such as those of patients with spasticity, has been traditionally performed by orthopedic surgery through an open approach. Continuous improvements in surgical techniques finally allowed using endoscopy to reach and treat the most common tendon disorders, ensuring better accuracy and greater respect of the anatomical structures.

9.3.1 Anatomy

The iliopsoas tendon is composed of the union of the great psoas muscle and iliac. These two, clearly separated proximally, come together on the femur at the level of the lesser trochanter.

The strong tendon lies in its course on the anterior portion of the capsule, at about 2 o'clock. It is at this level that a bursa can be found, which often have a direct communication with the coxofemoral joint.

The iliopsoas tendon injuries are frequently associated with pain, click, and strength reduction. This clinical condition, along with the anatomical and pathological findings described, is called internal snapping hip. It is noted that the snapping is present asymptomatically in approximately 10% of the population, and those most frequently affected by this clinical syndrome are athletes, especially dancer [11–13].

9.3.2 Diagnosis

The diagnosis of internal snapping hip is essentially clinical. With the hip in extension, the tendon is located medial to the femoral head. During hip flexion, the tendon moves laterally passing the front portion of the femoral head. The presence of a painful shutter during such maneuver together with the iliopsoas-positive test (flexion of the hip against resistance) confirms the diagnosis. The snap is never visible, sometimes may not be audible, but is palpable at the level of the reported area of pain and apprehension always associated with a response by the patient [14–16].

The iliopsoas muscle contributes to pelvic stability and is involved constantly in most sports activities. One reason why the muscle is at risk could be that its workload includes both eccentric and concentric work. When the injury is iliopsoas related, the pain is localized anterior of the proximal thigh, more laterally in the groin, may radiate to the anterior aspect of the femur, and sometimes produces some lower abdominal pain lateral to the rectus abdominis. Frequently, the muscle is also tight, and palpation just distal to the inguinal ligament is often painful. Palpation is performed above the inguinal ligament and lateral to the rectus abdominis. The iliopsoas can also be palpated in the area just below the inguinal ligament lateral to the femoral artery and medial to the sartorius muscle, the only area where the iliopsoas is directly palpable. The Thomas test should be performed to assess the tightness of the iliopsoas and to ascertain whether passive stretching is painful. Incomplete extension of the hip when performing the Thomas test is a sign of a tight iliopsoas muscle. Pressure by the examiner's hand to extend the hip further is a test for pain on passive stretching. The abovementioned tests for the iliopsoas were all found to be reproducible. Muscle weakness and pain when flexing the hip joint against resistance at 90° is often found. Sitting with the legs stretched and then elevating the heels might result in pain since the only active hip flexor in this position is the iliopsoas. This is the Ludloff's sign.

9.3.3 Imaging

Radiological study for these patients is essential to rule out any underlying conditions. Radiographs should be requested to include an anteroposterior, axial (or frog-leg), and Dunn at 90° to rule out a possible femoral-acetabular impingement. Ultrasound is a very valuable tool for dynamic evaluation of the tendon and can be taken into account in case of a possible infiltrative treatment, but its ability to diagnose shooting is currently insufficient.

MRI is usually used in case of diagnostic doubts or to better study the hip in order to rule out any intra-articular pathologies that might get in the differential diagnosis of hip internal shutter [16].

9.3.4 Conservative Treatment

The conservative treatment, in agreement with the literature data, is considered the gold standard for patients suffering from this syndrome. The therapeutic approach consists in reduction of sports activity, physical therapy, and anti-inflammatory. The rehabilitation program includes the strengthening of the muscles of the pelvic girdle, stretching and strengthening of the iliopsoas, assisted mobilization, and exercises to reduce the lumbar lordosis. After a period of about 3 weeks, in case the symptoms persist, it is possible to perform ultrasound-guided infiltration with local anesthetic and corticosteroids at the level of the bursa and in the peri-tendon area. Endoscopic treatment may be indicated if, for at least 3 months of treatment, the symptoms have not improved or the patient has reduced the activity level [17].

9.3.5 Surgical Technique

The endoscopic surgical technique for the treatment of the iliopsoas tendon was described for the first time by T. Byrd and subsequently revisited by several authors [18–20]. We can recognize three different anatomical areas where the release can be performed: the lesser trochanter, the peri-orbicular region, and the perilabral region (trans-capsular technique) (Picture 9.4).

The release to the lesser trochanter level is performed with the support of fluoroscopy. It guides both the optics and the instrument positioning, which converge at the level of the lesser trochanter; once isolated, a bursectomy and a complete detachment of all fibers are performed (Picture 9.5).

The release technique at the level of the periorbicular region is performed without traction and with the use of an accessory portal. Through the anterior portal, a capsulotomy of about 2 cm is performed, thus allowing a partial release of the tendon (Picture 9.6).

The release level of the perilabral region can be achieved with the use of standard portals used for access to the central compartment structures; we routinely use the anterolateral and mid-anterior portals. With leg still in traction, a trans-capsular window at about 2–3 o'clock is performed; at this level the tendon is located just above the joint capsule and can be easily recognized for its pearly color. The release is achieved through radiofrequencies going medial-lateral to protect the femoral nerve branches (Picture 9.7).

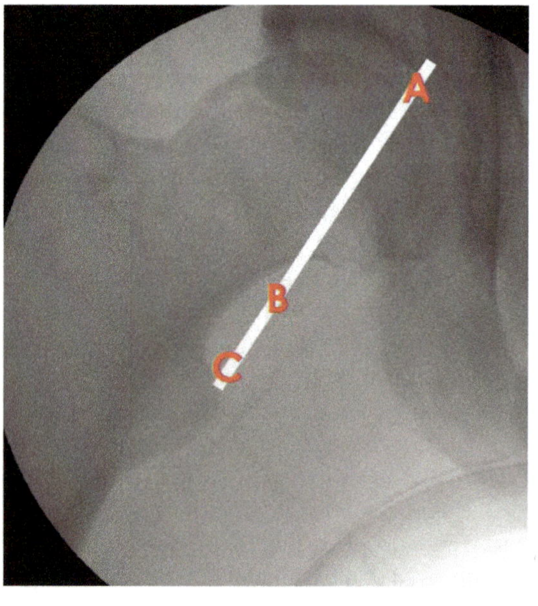

Picture 9.4 Different levels where a IP release can be performed. (*A*) Central compartment. (*B*) Peripheral compartment. (*C*) Lesser trochanter

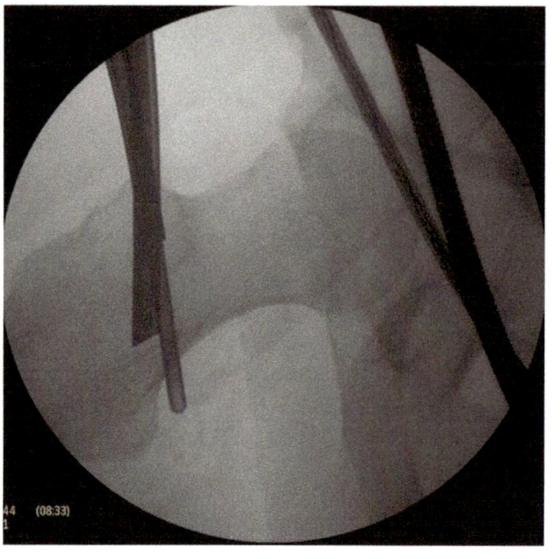

Picture 9.5 Lesser trochanter IP complete release

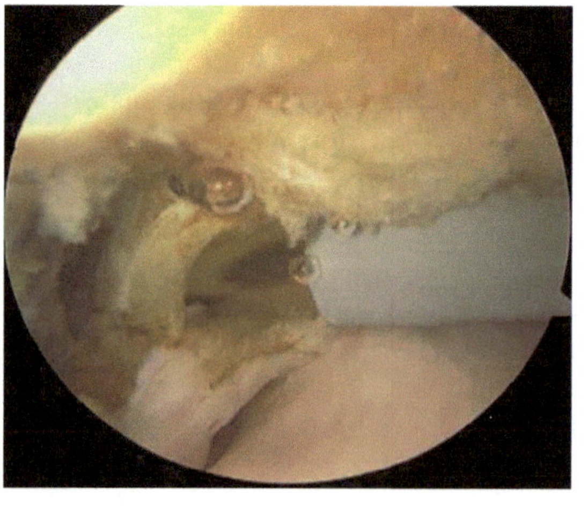

Picture 9.6 Peripheral compartment partial release

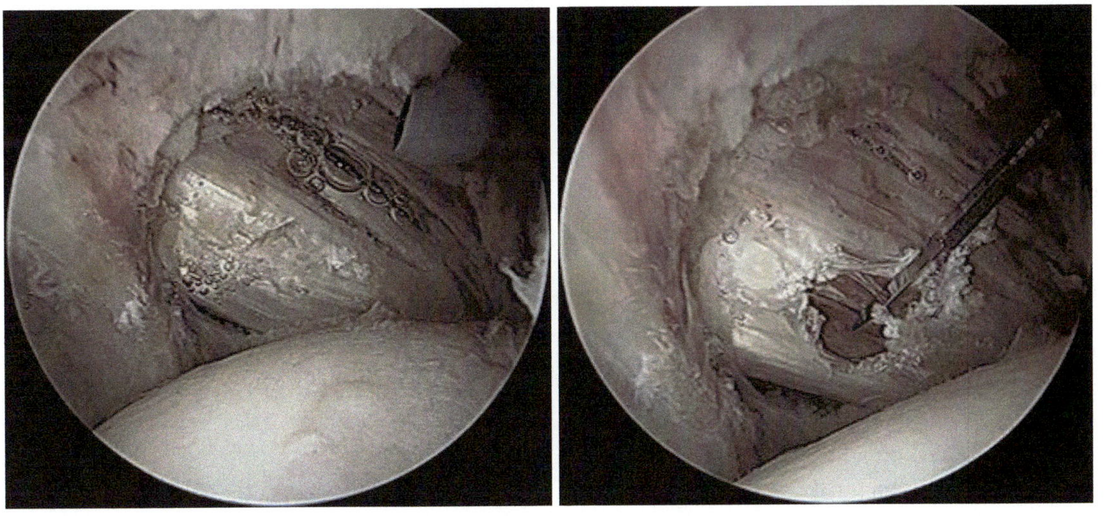

Picture 9.7 Central compartment IP partial release

Our choice to perform the trans-central capsular technique is based on these considerations:

- To achieve an elongation anatomically closer to the intra-articular structures
- To preserve the distal insertion at the level of the lesser trochanter
- To avoid the use of accessories portals to reach the tendon

9.3.6 Postoperative Rehabilitation

Patients undergoing an iliopsoas tendon release follow a rehabilitation protocol which prohibits the strengthening exercises of the hip flexors for the first 6 weeks, in order to promote muscle recovery of the bending strength. Strengthening exercises of the quadriceps and abductors may instead start from the first postoperative day. Stretching is recommended to prevent the formation of adhesions and capsular contracture.

In case of treatment of intra-articular pathologies or the peripheral space, to the information already given, those specific according to performed surgical gestures will be added.

9.3.7 Outcome

The results of arthroscopic treatment for the pathology of the iliopsoas tendon were evaluated by different studies, and different release techniques used seem to give the same results in clinical terms. The arthroscopic treatment, compared to the traditional open technique, combines a less invasive approach and respect of adjacent structures, with the possibility of intra-articular and peritrochanteric space treatment pathology that may be present (labrum tears, FAI etc.) [21, 22].

Ilizaliturri et al. have performed various studies on release techniques, from the one at the level of the lesser trochanter to the trans-capsular at the level of the acetabular margin. Their latest work includes a comparative evaluation of the two techniques, demonstrating broad equivalence in terms of clinical results for both [23]. Wettstein et al. have reported the trans-capsular technique results of nine patients showing a 100% success rate with a minimum of 3-month follow-up [24, 25].

Byrd reported his results with the release at the lesser trochanter, with resolution of symptoms in 100% of cases. The same author also indicates the presence of intra-articular lesions in 50% of treated subjects [19].

In one of his works, Ilizaliturri describes as the trans-capsular release technique may have a higher risk of recurrence, but neither in literature nor in our experience this possible complication has been clinically relevant; there is instead an evident security, simplicity, and easy reproducibility of the trans-capsular proximal technique in any condition compared to other endoscopic techniques [24, 25].

Our series on professional athletes suffering from internal snapping hip is at this time of 21 cases, all treated with a trans-capsular proximal technique and for which we are executing a strength test comparing hind limb flexion, in order to assess flexion strength. The results at 3 months do not show a significant strength loss compared to the contralateral, reinforcing the idea that the partial release is capable of resolving the pathology, leaving the ability of muscle response, particularly important especially in athletes.

References

1. Allen WC, Cope R. Coxa saltans: the snapping hip revisited. J Am Acad Orthop Surg. 1995;3:303–8.
2. Brignall CG, Brown RM, Stainsby GD. Fibrosis of the gluteus maximus as a cause of snapping hip. J Bone Joint Surg Am. 1993;75-A:909–10.
3. Zoltan DJ, Clancy WG, Keene JS. A new operative approach to snapping hip and refractory trochanteric bursitis in athletes. Am J Sports Med. 1986;14:201–4.
4. Faraj AA, Moulton A, Sirivastava VM. Snapping iliotibial band. Report of ten cases and review of the literature. Acta Orthop Belg. 2001;67:19–23.
5. Farr DO, Selesnick H, Janecki C, Cordas D. Arthroscopic bursectomy with concomitant ilio-tibial band release for the treatment of recalcitrant trochanteric bursitis. Arthroscopy. 2007;23:905.e1–5.
6. Ilizaliturri Jr VM, Martinez-Escalante FA, Chaidez PA, Camacho-Galindo J. Endoscopic iliotibial band release for external snapping hip syndrome. Arthroscopy. 2006;22:505–10.
7. Ilizzaliturri Jr VM, Camacho-Galindo J. Endoscopic treatment of snapping hips, iliotibilal band and ileopsoas tendon. Sport Med Arthrosc. 2010;18:120–7.
8. Glick JM, Sampson TG, Gordon RB, Behr JT, Schmidt E. Arthroscopy by lateral approach. Arthroscopy. 1987;3(1):4–12.
9. Polesello GC, Queiroz MC, Domb BG, Ono NK, Honda EK. Surgical technique: Endoscopic gluteus maximus tendon release for external snapping hip syndrome. Clin Orthop Relat Res. 2013;471(8):2471–6.
10. Zini R, Munegato D, De Benedetto M, Carraro A, Bigoni M. Endoscopic iliotibial band release in snapping hip. Hip Int. 2013;23(2):225–32.
11. Alpert JM, Kozanek M, Li G, Kelly BT, Asnis PD. Cross-sectional analysis of the iliopsoas tendon and its relationship to the acetabular labrum: an anatomic study. Am J Sports Med. 2009;37:1594–8.
12. Polster JM, Elgabaly M, Lee H, Klika A, Drake R, Barsoum W. MRI and gross anatomy of the iliopsoas tendon complex. Skeletal Radiol. 2008;37:55–8.
13. Winston P, Awan R, Cassidy JD, Bleakney RK. Clinical examination and ultrasound of self-reported snapping hip syndrome in elite ballet dancers. Am J Sports Med. 2007;35(1):118–26.
14. Register B, Pennock AT, Ho CP, Strickland CD, Lawand A, Philippon MJ. Prevalence of abnormal hip findings in asymptomatic participants: a prospective, blinded study. Am J Sports Med. 2012;40:2720–4.
15. Tatu L, Parratte B, Vuillier F, Diop M, Monnier G. Descriptive anatomy of the femoral portion of the iliopsoas muscle: anatomical basis of anterior snapping of the hip. Surg Radiol Anat. 2001;23:371–4.
16. Wahl CJ, Warren RF, Adler RS, Hannafin JA, Hansen B. Internal coxa saltans (snapping hip) as a result of overtraining: report of 3 cases in professional athletes with a review of causes and the role of ultra-sound in early diagnosis and management. Am J Sports Med. 2004;32:1302–9.
17. Lee KS, Rosas HG, Phancao JP. Snapping hip: imaging and treatment. Semin Musculoskelet Radiol. 2013;17(3):286–94.
18. JWT B. Hip arthroscopy: evolving frontiers. Oper Tech Orthop. 2004;14:58–67.
19. Byrd JWT. Snapping hip. Oper Tech Sports Med. 2005;13:46–54.
20. Byrd JW. Evaluation and management of the snapping iliopsoas tendon. Instr Course Lect. 2006;55:347–55.
21. Gruen GS, Scioscia TN, Lowenstein JE. The surgical treatment of internal snapping hip. Am J Sports Med. 2002;30:607–13.
22. Jacobson T, Allen WC. Surgical correction of the snapping iliopsoas tendon. Am J Sports Med. 1990;18:470–4.
23. Ilizaliturri Jr VM, Buganza-Tepole M, Olivos-Meza A, Acuna M. Central compartment release versus lesser trochanter release of the iliopsoas tendon for the treatment of internal snapping hip: a comparative study. Arthroscopy. 2014;30(7):790–5.
24. Ilizaliturri Jr VM, Chaidez C, Villegas P, Briseño A, Camacho-Galindo J. Prospective randomized study of 2 different techniques for endoscopic iliopsoas tendon release in the treatment of internal snapping hip syndrome. Arthroscopy. 2009;25(2):159–63.
25. Ilizaliturri Jr VM, Villalobos Jr FE, Chaidez PA, Valero FS, Aguilera JM. Internal snapping hip syndrome: treatment by endoscopic release of the iliopsoas tendon. Arthroscopy. 2005;21(11):1375–80.

Part 3

Visceral Pathology

Inguinal Hernia and Other Types of Hernia: Diagnostic and Therapeutic Approach

10

Francesco Di Marzo

10.1 Epidemiology

Ninety-five percent of patients are male, with incidence variation from 11 per 10,000 person years (aged 16–24) to 200 per 10,000 person years (aged 75 or above) [3].

The lifetime risk of developing an inguinal hernia is 3% for women and 27% for men [1]. The incidence rises with age and is eight times higher in persons with a positive family history [7].

The following risk factors have been described [8]:

- Chronic obstructive pulmonary disease
- Cigarette smoking
- Low body-mass index
- Collagen diseases

Lateral, medial, and femoral hernias are anatomically distinct from one another and arise at different frequencies [6]. Indirect hernias are twice as common as direct ones; femoral hernias account for less than 5% of all inguinal hernias (more common in women). Inguinal hernias are more often on the right side than the left [5, 9].

Obturator hernias account for 0.07–1% of all hernias and 0.2–1.6% of all cases of mechanical obstruction of the small bowel [10]. They have the highest mortality rate of all abdominal wall hernias (13–40%) [11].

Spigelian hernia accounts for 1–2% of all hernias; supravesical hernia is very rare (with few cases described in literature) [12, 13].

10.2 Diagnosis

A complete anamnesis is the first step to a correct diagnosis, and the following points are crucial: groin swelling (right/left), nature and duration of complaints (pain), contralateral groin swelling, symptoms of incarceration, reducibility, and previous hernia operations.

Predisposing factors must be investigated: smoking, chronic obstructive pulmonary disease (COPD), abdominal aortic aneurysm, long-term heavy lifting work, positive family history, appendicectomy, prostatectomy, and peritoneal dialysis [14].

A reducible (with proper maneuver) groin swelling (above the inguinal ligament) is definitive evidence of an inguinal hernia and needs no further diagnostic evaluation with the exception of physical examination [2].

Physical examination of the groin is mandatory not only in patients referred for lower abdominal pain; it consists of inspection (ipsi- and contralateral swelling, operation scar, skin signs of incarceration), followed by palpation of the patient's groin in both standing and lying

F. Di Marzo, M.D.
Department of General Surgery, Versilia Hospital - Azienda Usl Toscana Nord Ovest,
Lido di Camaiore, Italy
e-mail: fdimarzo@yahoo.com

R. Zini et al. (eds.), *Groin Pain Syndrome*, DOI 10.1007/978-3-319-41624-3_10

position, including digital exploration of the inguinal canal as afterward described and digital rectal exam. An inguinal hernia can be distinguished from a scrotal hernia (with or without hydrocele by palpation, using transillumination if necessary).

Evaluation must include testicle and spermatic cord bilateral palpation, inguinal canal exploration made by a finger into the external inguinal ring (at rest and during Valsalva maneuver), and palpation of bone and muscular-aponeurotic landmarks (pubic crest and tubercle, pectineal line, superior pubis ramus, inferolateral and superomedial pillar, common rectus abdominis-longus adductor muscle aponeurosis). In a recent study, a standardized questionnaire was used to evaluate symptoms of patients with inguinal hernia and in a control group [15]. Sixty-nine percent had discomfort in the hernia itself and 66% in the groin, while 50% complained of increased peristalsis, without any difference between right-sided, left-sided, or bilateral hernias. Only 7% had no symptoms. Patients with a left-side hernia complained mainly of increased peristalsis and tenesmus, while patients with right-side complained of urinary problems. The preoperative symptoms and the severity of pain in the early postoperative period were both risk factors for chronic pain [16].

Any non-reducible inguinal mass needs further diagnostic evaluation, even in asymptomatic patients. A meta-analysis confirmed the utility of ultrasonography for this purpose, with 96.6% sensitivity, 84.8% specificity, and a positive predictive value of 92.6% [27]. In a study of 36 patients with occult hernias, magnetic resonance imaging was found to be superior to both ultrasonography and computerized tomography [17].

Dynamic sonography is useful (low cost, good diagnostic value, and availability), but can only be stated as a grade C recommendation (suboptimal quality of studies).

Differential diagnosis to be considered is femoral or incisional hernia, lymph node enlargement, aneurysm, saphenous varix, soft tissue tumor, abscess, genital anomalies (ectopic testicle), adductor tendinopathy, pubic osteitis, hip artrosis, iliopectineal bursitis, low back pain, and endometriosis (in female patient) [14].

10.3 Classification

The European Hernia Society (EHS) Board agreed on a new classification based on Aachen system [18].

EHS classification defines the location of hernia with L (lateral), M (medial), and F (femoral). The size of hernia is indicated with 1 (≤1 finger), 2 (1–2 fingers), and 3 (≥3 fingers). For two different hernias in the same patient, appropriate boxes in the table are ticked. In addition, P or R letter is encircled for a primary or recurrent hernia.

No matter which classification system is used, the type of hernia should be recorded according to intraoperative findings, describing each side separately and clearly for bilateral hernias.

10.3.1 Indications for Treatment

The goal of treatment is to resolve symptoms, improve quality of life, and prevent adverse events such as incarceration while avoiding postoperative complications.

Surgery can improve the quality of life of patients with symptomatic inguinal hernias [19], even if they are elderly [20].

In men with asymptomatic or minimally symptomatic inguinal hernia, consider conservative management (watchful waiting).

Watchful waiting is a safe and acceptable option for men with minimally symptomatic or asymptomatic inguinal hernias. It is very likely (>70% chance) that, in time, the symptoms will increase leading to surgical intervention (evidence level 1B).

It is recommended in minimally symptomatic or asymptomatic inguinal hernia in men to consider a watchful waiting strategy, especially when older or in the presence of major comorbidity (recommendation grade B) [21].

In case of incarcerated hernia without symptoms and signs of strangulation, it is possible to try reduction; otherwise, emergency surgery is mandatory.

For symptomatic inguinal hernia, elective surgery is the choice.

For female patient consider femoral hernia and preperitoneal (endoscopic) approach.

Primary inguinal hernias in women should be operated in all cases because of the possibility of a femoral hernia, which cannot be unambiguously diagnosed and is incarcerated in up to 30% of cases (evidence level 2, recommendation grade B) [4, 14, 21].

In case of recurrent inguinal hernia, the decision must be made individually, in consideration of previous technique (with or without a mesh), symptoms, and comorbidities.

10.4 Surgery

Surgical procedures for inguinal hernia repair could be classified according to technique (primary tissue repair with suture or mesh positioning), to anatomical approach (anterior or posterior), and to type of surgery (open or laparoscopy/endoscopy).

Minimally invasive procedures are always done through a posterior approach and with the use of a mesh; open, primary tissue repair procedures are performed through the classic anterior approach. Bassini, Shouldice, and Desarda are suturing techniques; the standard mesh technique through an anterior approach is Lichtenstein [22].

The "plug and patch" technique and the use of special mesh systems (prolene hernia system, PHS) in open procedures to cover both the anterior and the posterior surface have been recently reviewed.

PHS and plug and patch result in comparable outcome (recurrence and chronic pain) as the Lichtenstein technique (1–4-year follow-up) (level 1A).

PHS and plug and patch (mesh plug) can be considered as an alternative treatment for Lichtenstein inguinal hernia repair (grade B) [21].

According to a recent meta-analysis of open suture and open mesh techniques, the Shouldice repair is associated with a lower recurrence rate than other suture techniques—Bassini—(7% vs. 4.3%) [23], but the recurrence rate of suture techniques is four times higher than that of mesh techniques (4% vs. 0.9%).

In the guidelines, the open Lichtenstein and endoscopic inguinal hernia (long learning curve for endoscopic repair, especially TEP—level 2C) techniques are recommended as best evidence-based options for repair of a primary unilateral hernia provided the surgeon is sufficiently experienced in the specific procedure [21].

For recurrent hernias after conventional open repair, endoscopic inguinal hernia techniques result in less postoperative pain, faster recovery, and less chronic pain than the Lichtenstein technique (evidence level 1A). For the repair of recurrent hernias after conventional open repair, endoscopic inguinal hernia techniques are recommended (recommendation grade A).

Material-reduced meshes have some advantages in terms of chronic pain and foreign body sensation in the first year after open surgery. There is, however, no difference in the incidence of severe chronic pain (level 1B).

This advantage has not been shown in endoscopic repair.

The use of lightweight/material-reduced/large-pore (>1000 μm) meshes in open inguinal hernia repair is advised with caution for large direct hernias (grade B).

Traumatic mesh fixation (non-resorbable devices) in TEP (with heavyweight mesh) is unnecessary in most cases (level 1A).

There is possibly a short-term benefit (postoperative pain) of atraumatic mesh fixation in the Lichtenstein procedure and in endoscopic procedures (TAPP). It offers no benefit in terms of chronic pain (level 1B).

When using heavyweight meshes, traumatic mesh fixation in TEP endoscopic repair should be avoided with exception for some cases like large direct hernias (grade B).

Atraumatic mesh fixation in the Lichtenstein technique and in TAPP endoscopic repair can be used without increasing the recurrence rate at 1 year (grade B).

The first study on the use of the self-gripping Parietene ProGrip mesh (large-pore polypropylene with resorbable polylactic acid micro grips) showed less pain on the first postoperative day vs. the use of another large-pore polypropylene mesh without self-gripping capacity [24]. Atraumatic mesh fixation (glue, self-fixating mesh) is more expensive than standard fixation, although the operation time was shorter in the majority of the studies. All studies with at least 1-year follow-up showed no differences in recurrence rates.

As a brief summary of procedure related to specific indications:

For primary unilateral, mesh repair is recommended (Lichtenstein or endoscopic, only if expertise is available).
For primary bilateral, mesh repair is recommended (Lichtenstein or endoscopic).
For recurrent inguinal hernia, mesh repair is recommended, modifying technique in relation to previous procedure (if anterior consider open preperitoneal mesh or endoscopic approach only if expertise is present; if previously posterior consider anterior mesh—Lichtenstein) [21].

10.4.1 Femoral Hernia [25, 26]

Femoral hernias account for 2–4% of groin hernias, are more common in women, and are located lateral and inferior to the pubic tubercle and inferior to the inguinal ligament. The diagnosis can usually be made by thorough physical examination. Some patients complain of pain in the ipsilateral lower extremity, which is caused by pressure on the femoral nerve. The pain disappears with flexion of the thigh (Astley Cooper sign).

Due to the high risk of hernia complications and better outcomes for elective versus emergency hernia repair, long-standing femoral hernias that are asymptomatic may be considered for watchful waiting, while recently identified hernias should be considered for elective repair, even if asymptomatic.

The risk of mortality for emergency operations was significantly higher than for elective operations (OR 5.37, 95% CI 3.24–8.91), highlighting the importance of repairing femoral hernias in an elective setting.

10.4.2 Obturator Hernia [10, 11]

Signs and symptoms resulting from obturator hernias are often vague and nonspecific. There is rarely a palpable mass as in other common abdominal wall hernias and diagnostic imaging can often be inconclusive. A high suspicion for obturator hernia should be maintained when assessing a patient presenting with bowel obstruction and intermittent symptoms or medial thigh pain are present. A full assessment of the hernial orifices including screening for the Howship–Romberg (inner thigh pain on internal rotation of the hip) sign should not be overlooked.

Early CT scanning should be considered in cases where inguinal and femoral hernias have been ruled out by clinical examination. Open approaches are well documented with good hernial defect repair rates. Laparoscopic repair may, also, be appropriate resulting in less postoperative pain, fewer pulmonary complications, and shorter hospital stays.

10.4.3 Spigelian Hernia [12]

Spigelian hernia (1–2% of all hernias) is the swelling through a congenital or acquired defect in the Spigelian aponeurosis (transversus abdominis aponeurosis limited by the linea semilunaris laterally and the lateral edge of the rectus muscle medially). Mostly, they lie in the “Spigelian hernia belt,” a transverse 6-cm-wide zone above the bisiliac plane; lower hernias are rare and should be differentiated from direct inguinal or supravesical hernias. Although named after Adriaan van der Spieghel, he only described the semilunar line (linea Spigeli) in 1645. Josef Klinkosch in 1764 first defined the Spigelian hernia as a defect in the semilunar line.

10.4.4 Supravesical Hernia [13]

Supravesical hernia is exceptional. It involves a hernia between the median umbilical ligament and the medial umbilical ligament and is classified as two types: internal supravesical hernia and external supravesical hernia. They are often the cause of intestinal obstruction.

Technical and anatomical aspect of surgical procedures is discussed in chapter "Sports Hernia: A Comparison of the Different Surgical Techniques."

References

1. Kingsnorth A, LeBlanc K. Hernias: inguinal and incisional. Lancet. 2003;362:1561–71.
2. Rosenberg J, Bisgaard T, Kehlet H, et al. Danish Hernia Database recommendations for the management of inguinal and femoral hernia in adults. Dan Med Bull. 2011;58:C4243.
3. Devlin HB. Trends in hernia surgery in the land of Astley Cooper. In: Soper NJ, editor. Problems in general surgery, vol. 12. Philadelphia: Lippincott-Raven; 1995. p. 85–92.
4. Chow A, Purkayastha S, Athanasiou T, Tekkis P, Darzi A. Inguinal hernia. BMJ Clin Evid. 2007;4:1–20.
5. Bay-Nielsen M, Kehlet H, Strand L, et al. Quality assessment of 26,304 herniorrhaphies in Denmark: a prospective nationwide study. Lancet. 2001;358:1124.
6. Schumpelick V, Treutner KH, Arlt G. Inguinal hernia repair in adults. Lancet. 1994;344:375.
7. Fitzgibbons Jr RJ, Forse RA. Clinical practice. Groin hernias in adults. N Engl J Med. 2015;372(8):756–63.
8. Jorgensen E, Makki N, Shen L, CHen D C, Tian C, Eckalbar W L, Hinds D, Ahituv N, Avins A. A genome-wide association study identifies four novel susceptibility loci underlying inguinal hernia. Nature communications 6, 10130 (2015). doi:10.1038/ncomms10130.
9. Burcharth J, Pedersen M, Bisgaard T, Pedersen C, Rosenberg J. Nationwide prevalence of groin hernia repair. PLoS One. 2013;8(1):e54367.
10. Mantoo SK, Mak K, Tan TJ. Obturator hernia: diagnosis and treatment in the modern era. Singapore Med J. 2009;50(9):866.
11. De Clerq L, Coenegrachts K, Feryn T, Van Couter A, Vandevoorde P, Verstraete K. An elderly woman with obstructed obturator hernia: a less common variety of external abdominal hernia. J Belg Radiol-Belg Tijdschr Radiol. 2010;93:302–4.
12. Skandalakis PN, Zoras O, Skandalakis JE, Mirilas P. Spigelian hernia: surgical anatomy, embryology, and technique of repair. Am Surg. 2006;72(1):42–8.
13. Skandalakis JE, Gray SW, Burns WB, Sangmalee U, Sorg JL. Internal and external supravesical hernia. Am Surg. 1976;42(2):142–6.
14. Simons MP, Aufenacker T, Bay-Nielsen M, Bouillot JL, Campanelli G, Conze J, de Lange D, Fortelny R, Heikkinen T, Kingsnorth A, Kukleta J, Morales-Conde S, Nordin P, Schumpelick V, Smedberg S, Smietanski M, Weber G, Miserez M. European Hernia Society guidelines on the treatment of inguinal hernia in adult patients. Hernia. 2009;13(4):343–403.
15. Pérez Lara FJ, Del Rey Moreno A, Oliva Muñoz H. Do we really know the symptoms of inguinal hernia? Hernia. 2015;19(5):703–12.
16. Bansal VK, Misra MC, Babu D, Victor J, Kumar S, Sagar R, Rajeshwari S, Krishna A, Rewari V. A prospective, randomized comparison of long-term outcomes: chronic groin pain and quality of life following totally extraperitoneal (TEP) and transabdominal preperitoneal (TAPP) laparoscopic inguinal hernia repair. Surg Endosc. 2013;27(7):2373–82.
17. Miller J, Cho J, Michael MJ, Saouaf R, Towfigh S. Role of imaging in the diagnosis of occult hernias. JAMA Surg. 2014;149(10):1077–80.
18. Miserez M, Alexandre JH, Campanelli G, Corcione F, Cuccu-rullo D, Pascual MH, et al. The European hernia society groin hernia classification: simple and easy to remember. Hernia. 2007;11:113–6.
19. Magnusson J, Videhult P, Gustafsson U, Nygren J, Thorell A. Relationship between preoperative symptoms and improvement of quality of life in patients undergoing elective inguinal herniorrhaphy. Surgery. 2014;155(1):106–13.
20. Pierides G, Mattila K, Vironen J. Quality of life change in elderly patients undergoing open inguinal hernia repair. Hernia. 2013;17(6):729–36.
21. Miserez M, Peeters E, Aufenacker T, Bouillot JL, Campanelli G, Conze J, Fortelny R, Heikkinen T, Jorgensen LN, Kukleta J, Morales-Conde S, Nordin P, Schumpelick V, Smedberg S, Smietanski M, Weber G, Simons MP. Update with level 1 studies of the European Hernia Society guidelines on the treatment of inguinal hernia in adult patients. Hernia. 2014;18(2):151–63.
22. Berger D. Evidence-based hernia treatment in adults. Dtsch Arztebl Int. 2016;113(9):150–8. doi:10.3238/arztebl.2016.0150.
23. Amato B, Moja L, Panico S, Persico G, Rispoli C, Rocco N, Moschetti I. Shouldice technique versus other open techniques for inguinal hernia repair. Cochrane Database Syst Rev. 2012;4:CD001543.
24. Kapischke M, Schulze H, Caliebe A. Self-fixating mesh for the Lichtenstein procedure—a prestudy. Langenbecks Arch Surg. 2010;395(4):317–22. doi:10.1007/s00423-010-0597-2.
25. Mattei P, editor. Fundamentals of pediatric surgery. New York: Springer; 2011. doi:10.1007/978-1-4419-6643-8_69.
26. Dahlstrand U, Wollert S, Nordin P, et al. Emergency femoral hernia repair: a study based on a national register. Ann Surg. 2009;249:672.
27. Robinson A, Light D, Nice C. Meta-analysis of sonography in the diagnosis of inguinal hernias. J Ultrasound Med. 2013;32(2):339–46.

11 Urogenital Pathology

Massimo Cecchi

11.1 Introduction

Groin pain encompasses a large number of possible etiologies. Even after thorough work-up, the etiology for groin pain can be very difficult to elucidate. Although the most common cause of groin pain is muscle, tendon, or ligament strain, before making the diagnosis of groin pain syndrome, other urological causes must be addressed such as urinary bladder processes, renal colic, and testis and prostatic diseases (Table 11.1).

Urologic diagnosis is very common, and some patients with groin pain must obtain an urologic evaluation for differential diagnosis. Pelvic pain is one of the most frustrating areas of urology. It is important to take a thorough history of possible urologic, intra-abdominal, or joint-related conditions. A prostate examination is important to rule out prostatitis, as well as a physical examination of the scrotum to rule out genital diseases.

However some patients who require urologic evaluation may have muscle or tendon disease and need orthopedic evaluation.

Table 11.1 Urological causes of groin pain

– Prostatitis
– Epididymitis
– Spermatocele
– Orchitis
– Hydrocele
– Retractile testicle
– Testicular cancer
– Testicular torsion
– Varicocele
– Scrotal masses
– Urinary tract infection
– Kidney and ureteral stone

11.2 Prostatic Diseases

11.2.1 Prostatitis

Prostatitis is a label applied to symptom complexes involving perineal discomfort and voiding symptoms. Stamey [1] defined this spectrum of prostatic disease as a "wastebasket of clinical ignorance." The non-urological symptoms of prostatitis include pain in the area of the anus, bladder, abdomen, and groin and pain during ejaculation. Clinical and laboratory investigations during the last three decades have changed our understanding of these pathologies.

Prostatitis is the most common urologic diagnosis in men younger than age 50 years; chronic prostatitis most commonly affects men older than 50 years [2]. Estimations show that 50% of men

M. Cecchi
U.O.C. Urology, Versilia Hospital Lido di Camaiore, Lucca, Italy
e-mail: massimo.cecchi@uslnordovest.toscana.it

R. Zini et al. (eds.), *Groin Pain Syndrome*, DOI 10.1007/978-3-319-41624-3_11

Table 11.2 Prostatitis classification of the National Institutes of Health [3]

- Type I acute bacterial prostatitis
- Type II chronic bacterial prostatitis
- Type III chronic abacterial prostatitis (type IIIa inflammatory CPPS or type IIIb non inflammatory CPPS)
- Type IV asymptomatic inflammatory prostatitis

will experience symptoms of prostatitis at some time in their lives [1].

Prostatitis is an infection or inflammation of the prostate which may be acute or chronic. Prostatitis is associated with substantial cost and significant resource consumption.

Traditional classifications present significant limitation, and the NIH classification has now been recognized as the best system for clinical practice and research (Table 11.2).

Etiology may be infective (gram-negative, gram-positive, anaerobic, chlamydial infection), chemical-induced inflammation (noxious substances in the urine), immunologic alteration, psychological factors, pelvic floor muscle abnormalities, and painful bladder syndrome.

The infective agents in acute and chronic bacterial prostatitis are the same in type and incidence as those that cause urinary tract infection. Common uropathogenic strains of *Escherichia coli* are the most frequent causes of prostatitis, whereas other coliform bacteria (*Proteus*, *Klebsiella*, *Enterobacter*, *Morganella*, *Serratia*, *Citrobacter*) and pseudomonas are the less frequent causes [1]. The role of chlamydial etiology of prostate infection is both confusing and conflicting.

Gram-negative Enterobacteriaceae and enterococci are responsible for most cases of bacterial prostatitis.

Krieger and Riley [4] showed that only 8% of patients with CPPS tested positive for microorganism. However chronic abacterial prostatitis may, in certain cases, actually be due to an occult, chronic bacterial infection [5].

The pathogenesis of bacterial prostatitis is unclear. It is known, however, that the infective agent must first reach the prostate, colonized the gland, and then invade the secretory system. The ascending urethral route of infection is the most common route with reflux into the prostatic ducts, but lymphogenous or hematogenous routes are also possible. Intraprostatic urinary reflux occurs commonly and is an important cause of prostatic calculi [6].

Risk factors of prostatic infection include alterated prostatic host defense, dysfunctional voiding, intraprostatic ductal reflex, immunologic alterations, chemical-induced inflammation, pelvic floor muscle abnormalities, neuroendocrine mechanisms, psychological factors, and painful bladder syndrome.

Acute bacterial prostatitis presentation is usually dramatic with fever; low back, groin, and perineal pain; voiding dysfunction; and constitutional symptoms (malaise, arthralgia, and myalgia). The prostate is tender, tense, swollen, and painful. The degree of pain during prostatic palpation is variable. Urinalysis reveals pyuria and bacteriuria, and cultures will be positive.

In contrast clinical manifestations of chronic prostatitis are variable and frequently subtle.

The term of prostatodynia or chronic pelvic pain syndrome (CPPS) is used to identify unexplained chronic pelvic pain in men. CPSS is an inflammatory condition of unknown cause. It is a catchall category of convenience for a heterogeneous mixture of male patients with longstanding symptoms with no objective explanation and no satisfactory treatment.

Genetic predisposition, autoimmunity, and abnormal functioning of the nervous system may play a role in the development of CPPS. Psychological stress and depression may play a role with local production of cytokines in the pelvis with increase of CPPS inflammation.

Cytokines, which are produced by white blood cells, genetic predisposition, autoimmunity, and abnormal functioning of the nervous system have been individually identified as a culprit in the causation of CPPS. At least in some cases, they may adversely affect the suppression of NGF with CPPS flare-ups. Psychological stress and depression may lead to a flare of CPPS symptoms.

This pain is associated with irritative voiding symptoms and/or pain located in the groin, genitalia, or perineum. Pontari and Ruggeri reviewed the numerous pathophysiologic mechanisms implicated as the potential etiologies of CPPS and concluded that, although the causes of CPPS remain unknown, the symptoms seem to arise for a combination of psychological factors and immune, neurologic, and endocrine system dysfunction [7].

Symptoms are prostatodynia (discomfort in a pelvic pain distribution), irritative voiding disfunction, and pain (pelvic, perineal, lower back, or groin). The impact on health status is significant with greatly diminished quality of life.

Mandatory evaluation includes history taking and physical and digital rectal examination. Culture techniques ([8, 10] and PPMT [9]) are essential for prostate inflammation diagnosis and classification. Pressure-flow studies, cystoscopy, transrectal ultrasonography, and pelvic imaging are optional studies.

Pain is the primary symptom of CPPS; the second most important component is urinary function. Impact of symptoms and quality of life need to be evaluated in patients with CPPS.

The predominant symptom in CPPS is pain which may be localized in the perineum, suprapubic area, penis, and groin. Most men experience variable symptoms of irritative voiding dysfunction, pain, or discomfort perceived in a pelvic pain distribution (suprapubic, perineal, groin, low back). Chills and fever are unusual. Postejaculatory discomfort and hemospermia may be present. Rectal examination may be normal or aspecific (enlarged or indurated prostate).

The origin of prostatodynia is unknown, and the term may frequently be misleading. Musculoskeletal abnormalities have been implicated and are probably responsible for many of the symptoms.

Diagnosis of prostatitis is performed with history and physical examination, rectal examination, and culture techniques (urinalysis, urine culture, lower urinary tract localization test). Prostatic imaging (ultrasound or NMR) is not recommended for routine examination; however, it may be useful in complex cases or for differential diagnosis. Transrectal ultrasonography can be useful in diagnosing and draining prostatic abscesses and diagnosing obstructed seminal vesicles or prostatic cysts. Flow rate, residual urine determination, and urine cytology can be useful for urinary tract evaluation.

Optional tests are urethral swab for culture, pressure-flow studies, video urodynamics, pelvic imaging (CT, NMR), and cystoscopy.

11.3 Scrotal Diseases

The acute scrotum requires similar management principles of "acute abdomen":

- Patient history
- Physical examination
- Imaging studies

11.3.1 Testis and Epididymis Pathology

Orchitis is inflammation of the testis which is a relatively rare condition; most cases are referred to as epididymo-orchitis. Orchitis and epididymitis may be bacterial, nonbacterial, noninfectious, chronic orchitis or epididymitis, chronic orchialgia, or epididymalgia.

On gross examination, the testis involved is enlarged, congested, and tensed; often the epididymis cannot be distinguished from the testis by palpation.

Chronic orchitis and orchialgia are usually secondary to trauma or acute bacterial orchitis. Scrotum is usually normal, but the testis may be indurated and tender; pain may be serious with groin irradiation.

Epididymitis may follow severe physical strain; pain is usually severe and may radiate along the spermatic cord and even reach the groin and the flank. Chronic epididymitis may result from recurrent epididymitis or other cause. Epididymis is tender and swollen. Pain is localized to epididymis with irradiation to testis and spermatic cord. Urethral swab is useful for culture and sensitivity testing.

11.3.1.1 Spermatocele

Spermatocele is a painless cystic mass containing sperm; most spermatoceles are less than 1 cm in diameter and asymptomatic. Large spermatocele may cause groin pain. Diagnosis usually is made easily upon physical examination, but scrotal ultrasound may be useful if there is any doubt.

11.3.2 Varicocele

Varicocele consists of dilatation of the pampiniform plexus above the testis usually on the left side. Physical examination reveals a mass of dilated, tortuous veins lying posterior to and above the testis. In most patients, a varicocele is of no clinical significance; a small percentage have problems with infertility. It may extend up to the external inguinal ring and may cause groin pain during Valsalva maneuver or physical strain.

11.3.3 Hydrocele

Hydrocele consists of a collection of fluid within the tunica or processus vaginalis.

A hydrocele may develop secondary to testicular neoplasm, orchitis and epididymitis, and local injury. However chronic hydrocele is more common and is usually asymptomatic. Patients with hydroceles have scrotal enlargement, a sensation of heaviness. The patient may however complain of its bulk or weight and inguinal or groin pain. On examination the mass can be distended tensely; ultrasound examination may be useful for differential diagnosis.

11.3.4 Torsion of the Spermatic Cord

Torsion of the spermatic cord causes severe pain in one testicle, lower abdominal or groin pain, and nausea and vomiting. However it may be accompanied only by moderate scrotal swelling and little or no pain with difficult differential diagnosis.

11.4 Kidney Diseases

11.4.1 Kidney and Ureteral Stone

Kidney and ureteral stones are very common. Symptoms can vary from severe pain to no pain at all depending on stone characteristics, such as size, shape, and location of stone in the urinary tract.

A kidney stone may cause severe pain in the loin and the flank; a ureteral stone may cause pain in the groin or thigh. Other symptoms are nausea, vomiting, ileus, painful urination, fever, and blood in the urine.

Distal ureteral stone causes pain that tends to radiate into the groin or testicle in the male or labia majora in the female because the pain is referred from the ilioinguinal or genitofemoral nerves. Pain may be acute or recurrent or dull.

The noncontrast computed tomography (CT) scan is the cornerstone of initial radiographic assessment. Ultrasonography can often show the location, shape, and size of stone and dilation of the urinary tract. Noncontrast CT scans have become the imaging modality of choice; in some situations, renal ultrasonography or a contrast study such as intravenous pyelography (IVP) may be preferred. Contrast CT scan study may sometimes help clarify a difficult or confusing case, but contrast scans are usually indicated only during subsequent evaluation.

Urine and blood examination are useful to evaluate renal function and urinary tract infection.

Conclusions

Urogenital pathologies are very common in men with groin pain. Since the etiology of groin pain can be very difficult to elucidate, urologic evaluation is mandatory in most cases.

References

1. Stamey TA. Pathogenesis and treatment of urinary tract infections. Baltimore: Williams & Wilkins; 1980.
2. McNsughton CM, Fowler Jr FJ, Elliott DB, Albersten PC, Barry MJ. Diagnosis and treatment chronic prostatitis: do urologist use the four glass test? Urology. 2000;55:403.
3. Krieger JN, Nyberg LJ, Nickel JC. NIH consensus definition and classification of prostatitis. JAMA. 1999;282:236.
4. Krieger JN, Riley DE. Chronic prostatitis: Charlottesville to Seattle. J Urol. 2004;172:2557.
5. Rudick CN, Berry RE, Johnson JR. Uropathogenic escherichia coli induces chronic pelvic pain. Infect Immun. 2011;79:628.
6. Kirby RS, Lowe D, Bultitude MI, Shuttleworth KE. Intraprostatic urinary reflux. An aetiological factor in abacterial prostatitis. Br J Urol. 1982;54:729.
7. Pontari MA, Ruggieri MR. Mechanisms in prostatitis/chronic pain syndrome. J Urol. 2004;172:839.
8. Meares EM, Stamey TA. Bacteriologic localization patterns in bacterial prostatitis and urethritis. Invest Urol. 1968;5:492.
9. Nickel JC. Clinical evaluation of man with chronic prostatitis/chronic pelvic pain syndrome. Urology. 2003;60:20.
10. Meares EM. Prostatitis and related disorders. In: Campbell's urology. 5th ed. Philadelphia: W.B. Saunders; 1986.

Part 4

Tendon/Muscle Pathology

12 Sports Hernia: Diagnostic and Therapeutic Approach

Antonio Guglielmi

In the last years, the term sports hernia has been used in sports medicine to define one of the causes of groin pain syndrome [1, 2]. The absence of a hernia can sometimes confuse the true meaning of the term. The surgeon Jerry Gilmore in the 1990s described a pathology with muscular and nervous aetiology also called Gilmore's groin, groin strain or sportsman's groin, distinct from classical inguinal hernia that presents a visceral extrusion [3].

It refers to a group of lesions of the lower abdominal wall and in particular localized at the muscular, tendon and nerve structures of the inguinal canal.

It is a painful disorder that affects predominantly male athletes [4] and is common in sports that involve rapid acceleration accompanied by sudden changes in direction, cutting and kicking, sneezing, turning and twisting. Soccer players are the sportsmen in which we found the largest number of this type of lesions, with lower frequency in rugby and ice hockey players and in runners [5, 6].

12.1 Aetiopathogenesis

Numerous causes have been hypothesized, but it is very difficult to establish from which it originated. We have to consider the multifactorial genesis of these lesions. An imbalance between the weaker lower abdominal muscles and the stronger hip adductor muscles depends from the physical characteristics of the athlete or more frequently from the different typology of athletic preparation. The trainers tend to prefer to work on strengthening of the leg muscle than on trunk muscles. Currently, the diffusion of soccer in young people shows a growing incidence of this pathology as a result of incorrect training. The result is a great difference of force distribution in these areas. During trunk flexion and rotation, the oblique internal and external abdominal muscles, both inserted onto the pubic bone, exert an upward traction, while the hip adductor muscles also inserted onto the pubic bone exert a downward traction. The muscular simultaneous contraction, for example, during a contrast in the match, exert a traction on the pubic bone due to opposing forces.

The gender difference of the musculoskeletal anatomy in this area may explain why the sports hernia is more common in males than females. For example, there is a different insertion of the rectus abdominis in females which inserts onto the antero-superior part of the symphysis, while in men, there is a continuity of these ligaments, one for each medial half of rectus with the gracilis and fascia lata [7, 8].

The pubic instability due to an altered coordination of muscular synergy, necessary for efficacious control of hip dynamics, may be the cause of sports hernia. Muscular imbalance and

A. Guglielmi
Villa Stuart Sports Clinic, Rome, Italy
e-mail: antonioguglielmi3@gmail.com

R. Zini et al. (eds.), *Groin Pain Syndrome*, DOI 10.1007/978-3-319-41624-3_12

functional instability of the pubic symphysis lead to small tears in the muscle wall and in the aponeurotic fascia, resulting in entrapment of the inguinal nerves and with low frequency of the iliohypogastric nerves, thus causing pain typically associated with sports hernia. Gullmo [9] suggested that the underlying cause of chronic groin pain is related to repeated stretching of the peritoneum or of the sensory fibres of the ilioinguinal nerve along its course. The ilioinguinal nerve, because of its anatomic course and its length, is easily exposed to injury and can be compressed or entrapped by tears in the abdominal muscle fascia (Fig. 12.1).

Another anatomic structure responsible for pain is the genital branch of the genitofemoral nerve which can be compressed by bulging of the posterior wall of a weak or incompetent inguinal canal, including laceration of the transverse fascia with protrusion of the preperitoneal fat. This condition may be associated with increased size of the angle of convergence between the insertion of the rectus muscle along with the conjoint tendon and the inguinal ligament resulting in an enlarged external inguinal ring. In addition, an alteration of the confluence between this tendon and the adductor muscle may be present, observing a fibrosis of these structures in chronic forms above the pubic crest.

In current reports [10–12], there is a discussion about the biomechanical alterations determined from "femoroacetabular impingement (FAI)", in particular cam type, and development of sports hernia.

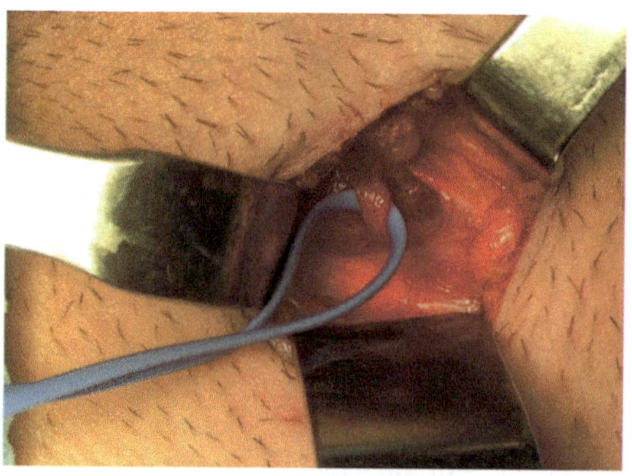

Fig. 12.1 Posterior wall of the inguinal canal and ilioinguinal nerve entrapped

The frequent association of these diseases should make us reflect on which disease starts first; in any event, there is an agreement on the treatment of both after a multidisciplinary evaluation.

12.2 Symptomatology and Physical Findings

Pain develops during or at the end of exercise or the day after the game. It is usually unilateral but occasionally bilateral and is located above the projection of the inguinal ligament to the lower lateral edge of the rectus abdominis muscle and often radiating to the scrotum (30%) and the proximal region of the thigh that is along the path of the adductor longus. Pain persists after a game and is accompanied by a difficulty in getting out of bed during the next 2 days.

The onset is insidious; it subsides with rest and reappears with sports activity return. The player is not able to kick with a combined movement of hip extra rotation. Pain is exacerbated by movement that involves sudden twisting and acceleration of the torso, by sneezing and coughing [13].

Clinical examination starts with the patient standing upright; occasionally, a slight bulge in the groin is observed. The simple palpation of the inguinal area above the pubic insertions of conjoint rectal muscle generally doesn't elicit pain.

The exploration through the scrotum (Fig. 12.2) reveals a dilated superficial inguinal ring due to the external oblique aponeurosis and/or conjoint tendon tear.

The thrust of the explorer finger will cause a deep and sharp pain enhanced by a crunch movement. If the patient coughs during this manoeuvre, you can feel a weak impulse of the posterior wall of the inguinal canal, and when a dislocated or entrapped nerve is found, the pain may continue even after the exploration (Fig. 12.1). This is a specific hallmark. To confirm the origin of pain, it is possible to make a lido test with the injection of local anaesthetic medially to the superior anterior iliac spine [14]. The immediate but temporary resolution of pain can help in differential diagnosis.

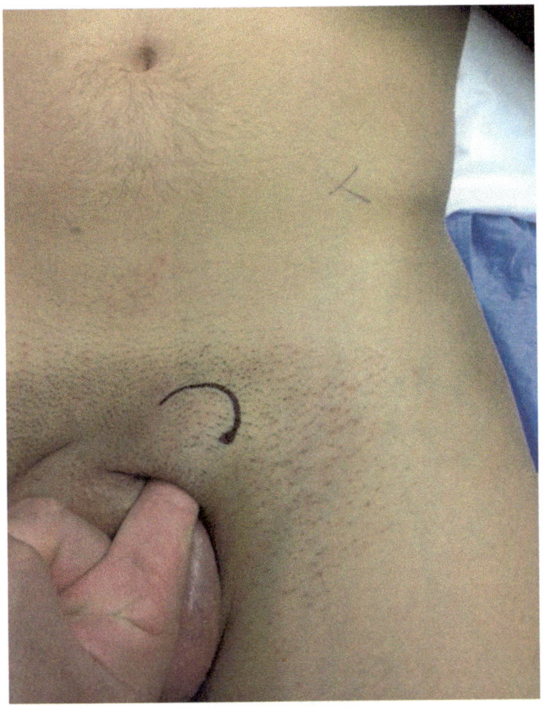

Fig. 12.2 Exploration through the scrotum with enlarged external inguinal ring

In order to exclude other associated diseases, you must carry out some semiological manoeuvres. The frequent association with cam FAI may be hypothesized with positive impingement test.

The pressure exerted on the pubic symphysis should suggest an osteitis. A positive squeeze test can highlight insertional lesions of the adductor muscles [15].

12.3 Instrumental Diagnosis

Dynamic ultrasound examination is certainly the most accessible but highly operator-dependent procedure [16]. An excellent knowledge of normal and pathologic groin anatomy is required. The examination may show macroscopic injuries of the conjoint tendon and of the aponeurosis of oblique external muscle. The lesions of the posterior wall of the inguinal canal are distinguished by a moderate and convex bulging associated with tear of the transversalis fascia and very small protrusions of peritoneal fat (Fig. 12.3). The picture is of an incipient direct inguinal hernia. The image capture is performed starting from the baseline condition, during a sit-up or when asked to contract the abdominal muscles (Valsalva). The local innervation and its course cannot be shown by ultrasound, but this is the area where we find entrapment or compression of the ilioinguinal nerve during open surgery. MRI is an examination of high diagnostic accuracy for muscle and tendon injuries; it allows us to define many associated diseases. It is required in second line to define more complex situations such as hip pathologies and pubis osteitis. The test is also made in basal condition and during Valsalva [17–19].

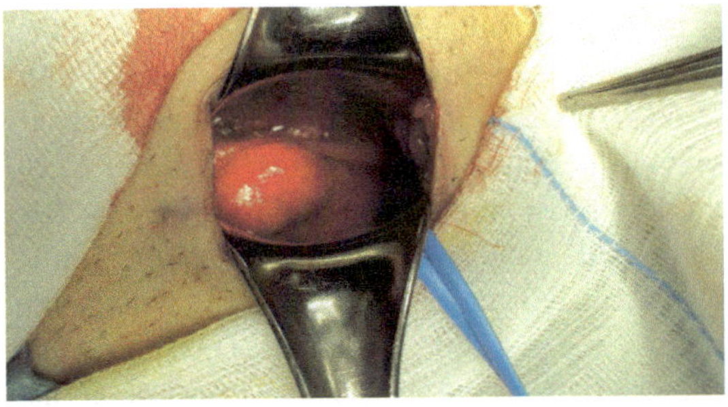

Fig. 12.3 Tears of the transversalis fascia and protrusion of peritoneal fat and nerve compression

12.4 Treatment

The athlete's history helps us to decide between a conservative and surgical treatment. The acute onset of symptoms suggests a conservative approach with anti-inflammatory therapy and rest. The specific physiotherapy required to redress the balance between the abdominal muscles, rectum and external oblique, internal and transverse and the most powerful adductor muscles will have to continue for a period of 8–12 weeks, depending on the severity of injury. The difference in the competitive level will have to be considered in choosing treatment. After 12 weeks, if there are still conditions for not playing a match, surgical treatment is called for. Although symptoms mainly refer to one side, it is preferable to operate on both sides in order to restore fair distribution of lines of force.

Surgical techniques proposed in the treatment derive from those used in traditional surgery for hernias [20, 21]. The open techniques with or without the mesh implant (Figs. 12.4 and 12.5) are designed to reconstruct the transversalis fascia, to reduce the angle between conjoint tendon and inguinal ligament and in strengthening the internal inguinal ring (Fig. 12.6). The association of a decompression or neurolysis of ilioinguinal and genitofemoral nerves and more rarely of iliohypogastric nerve depends on their anatomical position in the context of the injury observed. The observation is possible only with open repair techniques.

The most current video-assisted techniques (TAPP, TEP) [22, 23] with the placement of a polypropylene mesh report a faster return to sports activity. The possibility of positioning a prosthesis of a larger size allows us to stabilize different lesions of the posterior abdominal wall. Laparoscopic technique requires a longer learning curve than open surgery. Risks associated with general anaesthesia required for laparoscopic procedures and possible injury to the abdominal organs and vascular structures will have to be considered as in most cases we are dealing with very young and healthy patients.

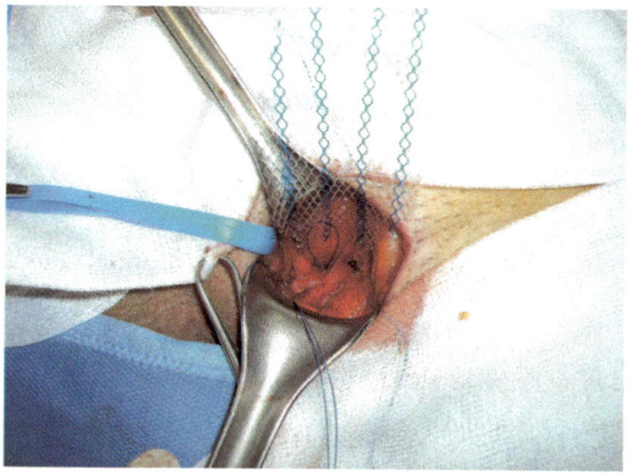

Fig. 12.4 Reconstruction of inguinal posterior wall, mesh implant

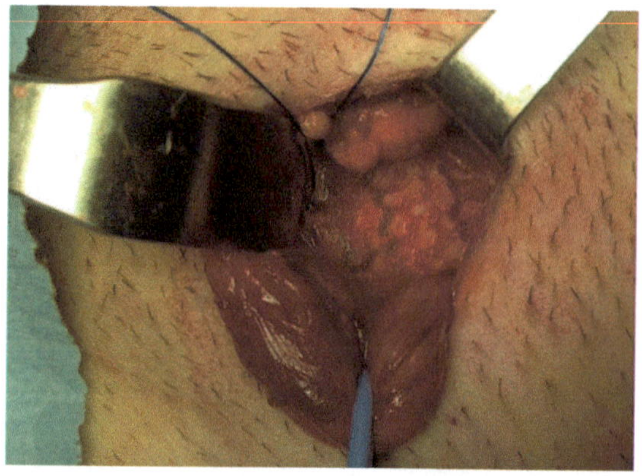

Fig. 12.5 Reconstruction of inguinal posterior wall, mesh implant

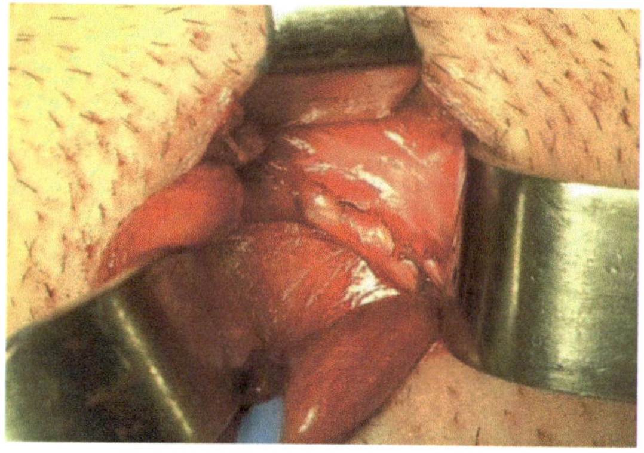

Fig. 12.6 Reconstruction of the conjoint tendon sutured to inguinal ligament

References

1. Weir A, Brukner P, Delahunt E, et al. Doha agreement meeting on terminology and definitions in groin pain in athletes. Br J Sports Med. 2015;49(12):768–74.
2. Sheen AJ, Jamdar S, Bhatti W. Calling for 'inguinal disruption' to be the term of choice for disorders of the inguinal ring: connecting Manchester and Doha. Br J Sports Med. 2016;50(7):442.
3. Gilmore J. Groin pain in the soccer athlete: fact, fiction, and treatment. Clin Sports Med. 1998;17:787–93.
4. Karlsson MK, Dahan R, Magnusson H, et al. Groin pain and soccer players: male versus female occurrence. J Sports Med Phys Fitness. 2014;54(4):487–93.
5. Paajanen H, Ristolainen L, Turunen H, et al. Prevalence and etiological factors of sport-related groin injuries in

top-level soccer compared to non-contact sports. Arch Orthop Trauma Surg. 2011;131(2):261–6.
6. Meyers WC, McKechnie A, Philippon MJ, et al. Experience with "sports hernia" spanning two decades. Ann Surg. 2008;248(4):656–65.
7. Bisciotti GN, Auci A, Di Marzo F, et al. Groin pain syndrome: an association of different pathologies and a case presentation. Muscles Ligaments Tendons J. 2015;5(3):214–22.
8. Robertson BA, Barker PJ, Fahrer M, Schache AG. The anatomy of the pubic region revisited: implications for the pathogenesis and clinical management of chronic groin pain in athletes. Sports Med. 2009;39(3):225–34.
9. Gullmo A. Herniography. The diagnosis of a hernia in the groin and incompetence of the pouch of Douglas and pelvic floor. Acta Radiol Suppl. 1980;361:1–76.
10. Strosberg DS, Ellis TJ, Renton DB. The role of femoroacetabular impingement in core muscle injury/athletic pubalgia: diagnosis and management. Front Surg. 2016;3:6.
11. Munegato D, Bigoni M, Gridavilla G, et al. Sports hernia and femoroacetabular impingement in athletes: a systematic review. World J Clin Cases. 2015;3(9):823–30.
12. Ross JR, Stone RM, Larson CM. Core muscle injury/sports hernia/athletic pubalgia, and femoroacetabular impingement. Sports Med Arthrosc. 2015;23(4):213–20.
13. Serner A, Tol JL, Jomaah N, et al. Diagnosis of acute groin injuries: a prospective study of 110 athletes. Am J Sports Med. 2015;43(8):1857–64.
14. Kulacoglu H, Ozyaylali I, Kunduracioglu B, et al. The value of anterior inguinal exploration with local anesthesia for better diagnosis of chronic groin pain in soccer players. Clin J Sport Med. 2011;21(5):456–9.
15. Weir A, Jansen J, van Keulen J, et al. Short and midterm results of a comprehensive treatment program for longstanding adductor-related groin pain in athletes: a case series. Phys Ther Sport. 2010;11(3):99–103.
16. Ostrom E, Joseph A. The use of musculoskeletal ultrasound for the diagnosis of groin and hip pain in athletes. Curr Sports Med Rep. 2016;15(2):86–90.
17. Yang DC, Nam KY, Kwon BS, Park JW, Ryu KH, Lee HJ, Sim GJ. Diagnosis of groin pain associated with sports hernia using dynamic ultrasound and physical examination: a case report. Ann Rehabil Med. 2015;39(6):1038–41.
18. Davies AG, Clarke AW, Gilmore J, et al. Review: imaging of groin pain in the athlete. Skelet Radiol. 2010;39(7):629–44.
19. Falvey ÉC, King E, Kinsella S, et al. Athletic groin pain (part 1): a prospective anatomical diagnosis of 382 patients-clinical findings, MRI findings and patient-reported outcome measures at baseline. Br J Sports Med. 2016;50(7):423–30.
20. Muschaweck U, Berger LM. Sportsmen's groin—diagnostic approach and treatment with minimal repair technique: a single-center uncontrolled clinical review. Sports Health. 2010;2(3):216–21.
21. Economopoulos KJ, Milewski MD, Hanks JB, et al. Sport hernia treatment: modified Bassini versus minimal repair. Sport Health. 2013;5(5):463–9.
22. Paajanen H, Montgomery A, Simon T, Sheen AJ. Systematic review: laparoscopic treatment of long-standing groin pain in athletes. Br J Sports Med. 2015;49(12):814–8.
23. Voorbrood CE, Goedhart E, Verleisdonk EJ, et al. Endoscopic totally extraperitoneal (TEP) hernia repair for inguinal disruption (Sportsman's hernia): rationale and design of a observational cohort study (TEP-ID-study). BMJ Open. 2016;6(1):e010014.

13 Sports Hernia: A Comparison of the Different Surgical Techniques

Francesco Di Marzo

Multidisciplinary evaluation (orthopedist, radiologist, surgeon, physiotherapist, sport physiologist) is mandatory to make diagnosis with confidence and to manage the whole treatment pathway (from first physical examination to "return to play"), avoiding recurrence, failure, and partial resolution [1, 2].

The level of evidence for operative treatment of sports hernia is poorly known, and a recent literature review shows a low level of study quality [3].

Surgery offers a wide range of procedural options and is part of the groin pain syndrome treatment, but considering it as the final step is a conceptual error.

This overview focuses on current surgical techniques, analyzing anatomical and step-by-step technical aspects.

Classification of current techniques is extremely important to define common points and to evaluate any difference:

Open technique without mesh positioning

Open all-suture repair
Minimal repair technique

Open technique with anterior onlay mesh positioning

Open with anterior mesh repair and combined adductor release if needed

Laparoscopic technique with posterior mesh positioning

Transabdominal preperitoneal
Totally extraperitoneal
Laparoscopic inguinal ligament release

At a glance, it's easy to establish two different ways in classifying all the procedures; the first is to consider how to approach the transversalis fascia:

1. Open technique through a skin/subcutaneous/aponeurosis incision
2. Videolaparoscopic/endoscopic technique through multiple micro-incisions, trocar positioning, and peritoneal flap creation (TAPP) or extraperitoneal space dissection (TEP)

The second takes into account the type of repair:

1. Primary tissue repair with suture
2. Mesh repair

The surgical strategy in laparoscopic technique is strictly tied to the use of a mesh in order to ensure a complete support to musculotendinous architecture of posterior wall without any correlation to nerves or inflammatory tissue (in the pubic area).

The open technique seems to be more flexible, allowing to choose a primary or a mesh repair combined with selective neurectomy if needed.

F. Di Marzo, M.D.
Department of General Surgery,
Versilia Hospital - Azienda Usl Toscana Nord Ovest,
Lido di Camaiore, Italy
e-mail: fdimarzo@yahoo.com

R. Zini et al. (eds.), *Groin Pain Syndrome*, DOI 10.1007/978-3-319-41624-3_13

13.1 Open All-Suture Repair

13.1.1 Anatomical Background

The rectus abdominis and adductor longus pull against the "pubic joint" with consequent weakness of posterior wall of the inguinal canal caused by force impairment (adductor's strength overcomes rectus), without any nerve entrapment [4, 5]. Sports hernia is treated as a muscular problem with a reinforcement of distal insertion of rectus abdominis and inguinal canal posterior wall, broadening the insertional surface area (including the rectus, internal oblique, and transversus abdominis via the conjoint tendon) [6].

13.1.2 Surgical Technique

The procedure is like a standard open inguinal hernia repair from skin incision to spermatic cord isolation and dissection exposing the posterior wall of inguinal canal.

The exploration of conjoint tendon, rectus abdominis insertion, and transversalis fascia is crucial to identify any defect (injury or medial/cranial retraction of rectus abdominis tendinous portion, conjoint tendon's tears, bulging of transversalis fascia).

Exposition of pubic tubercle and Cooper's ligament is necessary to provide a good reconstruction in which the first two stitches are crucial:

1. From lateral tendinous edge of the rectus abdominis to the roughed periosteum of pubic tubercle
2. From rectus edge (a little bit proximal to the first stitch) to the tubercle and the upper part of Cooper's ligament

The next stitch bites Cooper's ligament, lateral edge of the pubic tubercle, and reflected part of the ilioinguinal ligament. Additional sutures are placed as a Bassini repair to bring the conjoined tendon down to the ilioinguinal ligament (reflected part).

The rest of the procedure is, again, like a standard inguinal hernia repair.

In selected cases, an adductor tendon release is combined (contracted or overdeveloped adductor musculature).

13.1.3 Results

Litwin et al. reported good efficacy of the open, all-suture repair technique after evaluating 153 patients over a 6-year period from 2007 through January 2013.

Patient-reported outcome measure was used and showed positive results: physical health scores were improved by 15.5%, mental health scores were improved by 7.8%, and Tegner activity level scores were improved by 37.8%. A low rate of ipsilateral revision surgery and/or hip arthroscopy was reported [6].

13.2 Minimal Repair Technique

13.2.1 Anatomical Background

Muscles and aponeurosis of groin region interact as sliding planes [7].

Transversalis fascia weakness in the inguinal canal posterior wall with "bulging" creates a compression of genital branch of the genitofemoral nerve. The nerve distribution to the inner thigh or scrotum explains the irritation symptoms in such area [7].

Impaired force interaction between the rectus abdominis and adductor (due to overuse, typical of pro sport athletes) causes a chronic pro-inflammatory state in the pubic area, acting directly and mechanically on sensory nerves.

Realignment of muscles, inguinal canal posterior wall reinforcement, and simultaneous nerve decompression are the three main steps of the procedure.

13.2.2 Surgical Technique [7]

The minimal repair technique was developed by Dr. Ulrike Muschaweck, and it was first presented 14 years ago.

A 4-cm incision is made parallel to the inguinal ligament, more medial than for a standard inguinal hernia repair. A sharp incision is performed, and electrocautery dissects through the subcutaneous tissue and Scarpa's fascia. The external inguinal ring is identified and evaluated (looking for edema or fiber enlargement/attenuation). After the division of external oblique fibers, the ilioinguinal nerve is identified and preserved. The spermatic cord is retracted to expose the floor of the inguinal canal and transversalis fascia (an eventual hernia sac should be identified and treated).

It's important to use a Valsalva maneuver (direct or indirect) to clearly define a "bulging" in the floor. Careful evaluation of the genital branch of the genitofemoral nerve in this area can then be performed.

A lateral to medial incision of the transversalis is carried out, and a running suture (2/0 polypropylene) medial to lateral is performed between the iliopubic tract and the edge of the upper transversalis fascia; this first suture leaves a flap for the return suture. The running suture is continued to the lateral edge of the rectus muscle and anchored to it; then the flap is incorporated in the running suture laterally to the initial knot (imbricated suture line).

The final result is a complete obliteration of the bulging area.

The evaluation of the genital branch of the genitofemoral nerve in this area is mandatory; in case of inflammation or any other sign of involvement, it is excised.

To protect the ilioinguinal nerve from mechanical irritation, a collar made with internal oblique muscle fibers is prepared and placed around the nerve.

The rest of the procedure is, again, like a standard inguinal hernia repair.

13.2.3 Results

Both open techniques without mesh positioning were associated with equivalent results. The minimal repair technique has a short postoperative rehabilitation time which allows athletes to return to play very quickly (4.5 vs. 16.5 weeks) [7].

13.3 Open with Anterior Mesh Repair and Combined Adductor Release If Needed

13.3.1 Anatomical Background

Inguinal floor repair and stabilization with mesh (tension-free repair) avoid excessive tension (as in suture repair along the suture line) and its sequelae (postoperative chronic pain, recurrence) [8]. Alternative strategies are necessary in particular cases: young not completely grown male athletes, young female athletes (open primary tissue repair), and prior sports hernia or inguinal hernia open repair (laparoscopic approach) [9].

The choice to combine an adductor release procedure is related to tension reduction at its proximal insertion.

13.3.2 Surgical Technique

A standard incision and dissection through the subcutaneous tissue and Scarpa's fascia as for an inguinal hernia repair are performed. The external inguinal ring is identified, and the two pillars are evaluated (tension, dilation, fiber attenuation). After the division of external oblique fibers, the ilioinguinal nerve is identified and preserved. The spermatic cord is retracted to expose the floor of the inguinal canal and transversalis fascia (an eventual hernia sac should be identified and treated).

The floor is reconstructed by suturing a lightweight polypropylene mesh to the transversalis fascia and rectus sheath with single nonabsorbable stitches. The lateral edge is sutured to the inguinal ligament with a running polypropylene suture, the mesh is split, and the two ends are sutured together to the inguinal ligament. As in Lichtenstein repair, a deep inguinal ring reconstruction with the two tails of the mesh is performed: the goal is not the ring reconstruction itself but to allow a linear and clear adhesion of the mesh to the floor.

Also PTFE and biologic mesh could be used for repair [10–12].

The rest of the procedure is, again, like a standard inguinal hernia repair with some difference in nerve management.

The usual approach is to routinely identify the ilioinguinal nerve and to resect it, to avoid a source of postoperative pain, in selected cases (if it makes a turn through a slit in the external oblique rather than the external ring or if it runs the risk of being exposed to the mesh scarring process). In case of resection, it should be carried out 3–4 cm proximal to the internal ring and the stump allowed to retract into the muscle bed to minimize the risk of neuroma formation [9].

Also the role of adductor release is not clear in literature.

In case of adductor release procedure, with the thigh flexed and abducted, an incision (2/2.5 cm) over the upper adductor longus tendon is carried out, exposing the sheath, and a series of small cuts in the epimysium are made for about 3 cm [9].

13.3.3 Results

This technique shows good results in terms of return to play (more than 90% in different series), recurrence, and failure. A very low number of patients required subsequent hip surgery or developed hip-related pathology.

13.4 Laparoscopic Repair with Mesh (Transabdominal Preperitoneal, Totally Extraperitoneal)

13.4.1 Anatomical Background

A tear of the abdominal wall in the posterior inguinal canal or conjoined tendon, with or without bulging (like a medial hernia), and tissue damage or disruption of the posterior wall are analogous to an incipient direct inguinal hernia (with or without bulge) [5, 13].

The posterior positioning of the mesh, covering the area that will include the conjoint tendon and pubic bone, creates theoretically a stronger and wider support than a conventional anterior approach.

13.4.2 Surgical Technique

The approach is a standard laparoscopic inguinal hernia repair (TAPP). All laparoscopic/endoscopic procedures are performed under general anesthesia with the patient in a supine or slide Trendelenburg position and empty bladder. In TAPP technique, pneumoperitoneum is established, using an open access below or above (it depends on umbilical-pubis distance) the umbilicus to minimize the risk of Veress needle use (incidental visceral or vascular injury). A 10-mm port is placed with 30° camera; two 5-mm working ports are introduced for dissection on the transverse umbilical line at cross with the right and left midclavicular line. A lateral to medial peritoneal incision is carried out from two fingers medial to the anterior superior iliac spine to the medial umbilical ligament (obliterated umbilical artery). The dissection is strictly related to the transversalis fascia architecture: a two-layered structure. Superiorly the two layers are distinct, and inferiorly they insert onto the Cooper's ligament. The anterior layer is adherent to the rectus abdominis muscle, and the posterior layer lies in between the anterior and the peritoneum, dividing this space into an anterior (vascular space) and a posterior (avascular space of Bogros). The transversalis fascia does not provide significant strength and integrity to the groin; it bridges the space between the transversus abdominis arch superiorly and inguinal and Cooper's ligament inferiorly. Condensations of transversalis arch form three structure: interfoveolar ligament (fibers vertically oriented form the medial margin of internal inguinal ring; it has no significant strength), iliopubic tract (anterior to Cooper's and posterior to inguinal ligament, it is the superior boundary of the triangle of pain in which lie the nerves at risk of injury and entrapment), and iliopectineal arch (covering iliac muscle arches, it gives origin to a portion of the internal oblique and transversus abdominis muscle) [14]. The lateral dissection is carried out in the posterior space of Bogros and the medial between the anterior and the posterior layer very close to the rectus abdominis in order to expose the Cooper's ligament, the pubic tubercle, and the crucial area medial to epigastric vessels. A 15 × 10 cm mesh is chosen, allowing a large overlap on the injury of the posterior wall. The mesh is unfolded in the preperitoneal space and fixed onto the abdominal wall with tissue glue or without any fixation in case of self-gripping mesh. Peritoneal flap suture is usually performed by running suture (in recent report with barbed suture). Laparoscopic mesh positioning in TAPP usually follows the current principle of inguinal hernia surgery and is based on the physiological principle of diffusing the intra-abdominal pressure on the whole mesh in the disrupted groin area. This ensures immediate fixation of the mesh to the abdominal wall [16] with a complete covering of the crucial injured areas at the pubic symphysis complex and a firm support to the posterior wall of the inguinal canal.

In TEP, the preperitoneal space is approached through a small subumbilical skin incision, reaching the preperitoneal facial space using a blunt 10-mm port on the affected side through the anterior rectus fascia and a muscle split. It is possible to create the room for the procedure with balloon dilation or by using the camera together with gas insufflation (low cost). After dissection and visualization of standard anatomical landmarks, the retropubic area is covered with a 15 × 10 cm mesh, using no fixation or tissue glue [17].

13.4.3 Results

Athletes return to activity in 70–90 % of cases already after 3–4 weeks after both TAPP and TEP. Most frequent complications are superficial surgical site infection or minor bleeding. Recurrent groin pain may occur, but severe postoperative neuralgias have not been reported. In Ingoldby's series, 2 players out of 14 had prolonged neuralgia after TAPP which settled by 2 months [18]. In the study by Paajanen et al., 2 out of 30 operated patients (TEP) had recurrences of sports hernia (lateral hip pain with FAI, hip surgery) [19].

The TEP technique has the advantage of being performed without entering the abdominal cavity and avoiding potential risk of harming intra-abdominal structures but with a higher risk of vascular injury. The preperitoneal endoscopic technique results in less dominant scars than the open techniques.

In a recent meta-analysis, it was concluded that TEP has shorter recovery time but slightly higher intraoperative morbidity than TAPP [20].

There are no differences related to fixation if a self-gripping mesh is used.

Both open and laparoscopic repairs are good and show equivalent results. Laparoscopic approach may allow earlier return to full sport [21]. Endoscopic repair in athletes has some advantages: postoperative pain and wound complications are less frequent after laparoscopic surgery than in open surgery.

13.5 Inguinal Ligament Release + Mesh (Lloyd Release Procedure)

13.5.1 Anatomical Background

Inguinal ligament is the distal reflection of the external oblique muscle and should be defined as a tendon; it is attached uniformly onto the pubic tubercle, and the lower fibers are seen to merge with the lacunar ligament with an intact free edge at the medial wall of the femoral canal. In patient with groin pain, there are abnormalities (inguinal ligament enthesopathy) [21]. Biopsies taken from the inguinal ligament attachment reveal degradation of the normal collagen architecture with loss of nuclei, disruption of collagen fibers, and hemosiderin deposition. There is edema of the tissues and early mucoid degeneration and a paucity of inflammatory cells similar to the findings described in tennis elbow "angiofibroblastic tendinosis" [22]. Releasing of tension in the inguinal ligament attachment at the pubic bone results in patient's symptom improvement [21].

13.5.2 Surgical Technique [21]

The approach is a standard laparoscopic inguinal hernia repair (TAPP). The inguinal ligament is exposed creating a peritoneal flap and, by dividing the transversalis fascia, is then identified from the femoral canal to the external ring and divided at its attachment to the pubic tubercle. Any additional scar tissue is divided. The fascia overlying the pectineus and some fibers of the rectus sheath attachment may have to be released. The dissection allows exposition of the whole posterior inguinal region which is reinforced with standard mesh (15 × 12 cm) similar to that used in TAPP hernia repair.

13.5.3 Results

More than 90% of patients returned to normal sporting activity following surgery, with a median return to play of 4 weeks [21].

13.6 Key Points Summary

Nerve management. Irshad and colleagues [11] reported one or more tears in the external oblique aponeurosis within which branches of the ilioinguinal or iliohypogastric nerves exited in a series of pro-athletes with groin pain who underwent operative management. In a later study from the same group [12], some form of entrapment or tension on the nerve or its branches was felt to present in every case, but pathologic findings of nerve injury (axonal degeneration, perineural fibrosis) were identified in only less than one fourth of specimens.

Selective resection of the genital branch of genitofemoral nerve has been advocated by Muschaweck [7], but this approach has not been used by other groups.

It's important to remember that:

1. All nerves passing through the inguinal region are both sensitive and motor.
2. Communication between genitofemoral and ilioinguinal nerve is common; this results in overlap of sensory innervation, which can cause difficulties in preoperative definition of pain distribution and eventual entrapment.

13.7 Fixation

Fixing the mesh could be the origin of pain for direct nerve harm at the point of fixation, inflammatory response with mesh shrinkage, and traction in the mechanical fixation point resulting in chronic pain.

In laparoscopic procedures, there are four different methods for fixation: tacks, glue, no fixation (for normal mesh), and self-gripping mesh. In a recent meta-analysis, it was demonstrated that glue fixation reduced the risk for developing chronic pain in endoscopic hernia repair [17].

Conclusion

Surgical exploration and repair should be considered as first option when diagnosis is confirmed by a multidisciplinary team, to avoid conservative treatment failure and delayed recovery. A wide variety of open repair techniques are described to reinforce the posterior wall of the inguinal canal with or without a mesh, through open or laparoscopic/endoscopic approach. Combined procedures are inguinal ligament release, simultaneous percutaneous adductor tenotomy, excision of the ilioinguinal nerve, and resection of the compressed genital branch of the genitofemoral nerve. Femoroacetabular impingement is present in a high percentage of case and represents a source of discussion: is hip surgery necessary? Before or after inguinal surgery? The best management is yet to come in the form of a sequential surgery guided by patient symptoms and adaptation to overuse. Overuse is central in the development of GPS: we cannot modify the *primum movens*; we are all "treating the effect."

Laparoscopic approaches show a quicker return to sporting activity than with the open approach with a higher recurrence rate compared with an open mesh repair.

Therefore, it is still debatable which operative approach is more effective. Every group or center focused on GPS diagnosis and treatment is prone to offer only one procedure because of case load, specific surgical skill, scientific belief, and, obviously, good outcomes.

Three unpublished RCT will clarify some points about surgical treatment of sports hernia:

1. Total ExtraPeritoneal (TEP) Versus Open Minimal Suture Repair for Treatment of Sportsman's Hernia/Athletic Pubalgia: A Randomized Multi-center Trial (100 patients)
2. A Randomised, Blinded Study on Laparoscopic Mesh Reinforcement for Chronic Groin Pain (80 patients)
3. Total ExtraPeritoneal (TEP) Versus Open Minimal Suture Repair for Treatment of Sportsman's Hernia/Athletic Pubalgia: A Randomized Multi-center Trial (60 patients)

References

1. Bisciotti GN, Auci A, Di Marzo F, Quaglia R, Volpi P. Groin pain syndrome: an association of different pathologies and a case presentation. Muscles Ligaments Tendons J. 2015;5(3):214–22.
2. Bisciotti GN, Volpi P, Zini R, Di Marzo et al Groin Pain Syndrome Italian Consensus Conference on terminology, clinical evaluation and imaging assessment in groin pain in athlete BMJ Open Sport & Exercise Medicine 2016;2:e000142. doi: 10.1136/bmjsem-2016-000142.
3. Serner A, van Eijck CH, Beumer BR, Hölmich P, Weir A, de Vos R-J. Study quality on groin injury management remains low: a systematic review on treatment of groin pain in athletes. Br J Sports Med. 2015;49(12):813.
4. Meyers WC, Lanfranco A, Castellanos A. Surgical management of chronic lower abdominal and groin pain in high-performance athletes. Curr Sports Med Rep. 2002;1:301–5.
5. Meyers WC, McKechnie A, Philippon MJ, Horner MA, Zoga AC, Devon ON. Experience with "sports hernia" spanning two decades. Ann Surg. 2008;248:656–65.
6. Litwin DE, Sneider EB, McEnaney PM, Busconi BD. Athletic pubalgia (sports hernia). Clin Sports Med. 2011;30:417–34.
7. Muschaweck U, Berger L. Minimal repair technique of sportsmen's groin: an innovative open-suture repair to treat chronic inguinal pain. Hernia. 2010;14: 27–33.
8. Treadwell J, Tipton K, Oyesanmi O, Sun F, Schoelles K. Surgical Options for Inguinal Hernia: Comparative Effectiveness Review. Comparative Effectiveness Review No. 70. (Prepared by the ECRI Institute Evidence-based Practice Center under Contract No. 290-2007-10063.) AHRQ Publication No. 12-EHC 091-EF. Rockville, MD: Agency for Healthcare Research and Quality. August 2012. www.effectivehealthcare.ahrq.gov/reports/final.cfm.
9. Brunt LM, Barile R. My approach to athletic pubalgia. In: Byrd TW, editor. Operative hip arthroscopy. 3rd ed. New York: Springer; 2013. p. 55–65.
10. Joesting DR. Diagnosis and treatment of sportsman's hernia. Curr Sports Med Rep. 2002;1:121–4.
11. Irshad K, Feldman L, Lavoie C, Lacroix V, Mulder D, Brown R. Operative management of "hockey groin syndrome": 12 years experience in National Hockey League players. Surgery. 2001;130:759–66.
12. Brown R, Mascia A, Kinnear DG, Lacroix VJ, Feldman L, Mulder DS. An 18 year review of sports groin injuries in the elite hockey player: clinical presentation, new diagnostic imaging, treatment, and results. Clin J Sports Med. 2008;2008:221–6.
13. Gilmore OJA. Gilmore groin: ten years experience of groin disruption—a previously unsolved problem in sportsmen. Sport Med Soft Tissue Trauma. 1991;3:12–4.
14. Kingsnorth AN, et al. Embryology, anatomy and surgical applications of the pre peritoneal space. Surg Clin North Am. 2000;80(1):1–24.
15. Di Marzo F, Bisciotti GN, Mingarelli V, Gia L. Accepted abstract @EHS; 2016.
16. Bittner R, Arregui ME, Bisgaard T, et al. Guidelines for laparoscopic (TAPP) and endoscopic (TEP) treatment of inguinal hernia [International Endohernia Society (IEHS)]. Surg Endosc. 2011;25:2773–843.
17. Sajid MS, Ladwa N, Kalra L, et al. A meta-analysis examining the use of tacker mesh fixation versus glue mesh fixation in laparoscopic inguinal hernia repair. Am J Surg. 2013;206:103–11.
18. Ingoldby CJ. Laparoscopic and conventional repair of groin disruption in sportsmen. Br J Surg. 1997;84:213–5.
19. Paajanen H, Syvähuoko I, Airo I. Totally extraperitoneal endoscopic (TEP) treatment of sportsman's hernia. Surg Laparosc Endosc Percutan Tech. 2004;14:215–8.
20. Antoniou SA, Antoniou GA, Bartsch DK, et al. Transabdominal preperitoneal versus totally extraperitoneal repair of inguinal hernia: a meta-analysis of randomized studies. Am J Surg. 2013;206:245–52; pii:S0002-9610(13)00223-7.
21. Lloyd DM, Sutton CD, Altafa A, et al. Laparoscopic inguinal ligament tenotomy and mesh reinforcement of the anterior abdominal wall: a new approach for the management of chronic groin pain. Surg Laparosc Endosc Percutan Tech. 2008;18:363–8.
22. Nirschl RP, Pettrone FA. Tennis elbow. The surgical treatment of lateral epicondylitis. J Bone Joint Surg Am. 1979;61(6A):832–9.

14 Adductor and Upright Abdominal Tendinopathy

Gian Luigi Canata and Valentina Casale

14.1 Introduction

Insertional tendinopathy of the adductors and the rectus abdominis is common in male athletes, especially in soccer players [1]. Dysfunction of the adductor muscles represents the most common single cause of groin pain in athletes [2].

Insertional tendinopathies are one of the several causes of chronic groin pain. This condition varies among different sport activities, but it especially develops in the presence of repeated kicking and rapid changes of direction [3].

The exact incidence of this type of tendinopathies remains unknown because of the several different pathologies that might cause groin pain, even if some authors agree with a 2.5–3% incidence of rectus-adductor tendinopathy [4]. More than 70% of patients are males [5].

The wide variety of possible injuries in several anatomical structures and the high prevalence of "abnormal findings" in asymptomatic athletes contribute to the complexity of the "groin pain" issue in athletes [6].

The etiopathogenesis of insertional tendinopathies is multifactorial. It is usually related to repeated microtraumas and functional overuse, especially when physical activity requires sudden changes of direction, repeated acceleration and deceleration, and torsion and traction of abdominal and adductor muscles [1]. The most frequently involved sports are soccer, hockey, rugby, football, long-distance running, and basketball indeed [5].

A further predisposing factor is the lack of balance between a hypertonic adductor muscle and a hypotonic large flat muscular sheet of the abdomen, sometimes in addition to a hypertonic femoral quadriceps muscle [7].

The abdominal and paravertebral muscles act together to stabilize the pubic symphysis, especially during static or dynamic single-leg stance. The adductor muscles act as antagonists on the pubic symphysis, through an opposite traction on it [8].

The femoral triangle is located in the upper inner thigh and includes the femoral nerve, the femoral vessels, the sartorius, the iliopsoas, the pectineus, and the adductor longus muscles [9]. The adductor muscles are the adductor longus, the pectineus, and the gracilis in the superficial layer; the adductor brevis in the intermediate layer; and the adductor magnus in the deep layer. Two further examples of adductor muscles are the obturator externus and the quadratus femoris, though they are part of the gluteal region. The adductor longus is thought to be the most frequently injured adductor muscle: its origin and insertion do not guarantee mechanical advantages, making it more susceptible to traumatic lesions [10].

G.L. Canata, M.D. (✉) • V. Casale, M.D.
Centre of Sports Traumatology, Koelliker Hospital, Turin, Italy
e-mail: studio@ortosport.it

R. Zini et al. (eds.), *Groin Pain Syndrome*, DOI 10.1007/978-3-319-41624-3_14

This type of insertional tendinopathy involves both the adductor muscles (gracilis, pectineus, adductor longus, and adductor magnus) (Fig. 14.1) and the rectus abdominis muscle [11] (Fig. 14.2). Arthropathy of the pubic symphysis and insertional pubic area may also occur in advanced phases.

Several studies have addressed both intrinsic and extrinsic possible risk factors.

The main intrinsic factor consists of the strength imbalance between the adductor and abdominal muscles, but also reduced flexibility, important asymmetry or dysmetria between the lower limbs, lumbar hyperlordosis, and sacroiliac or hip arthropathy are listed as intrinsic risk factors for the development of adductor and abdominal tendinopathy [9]. Other extrinsic risk factors are incorrect athletic training, unfavorable conditions of the playground, and wrong sport equipment, above all footwear.

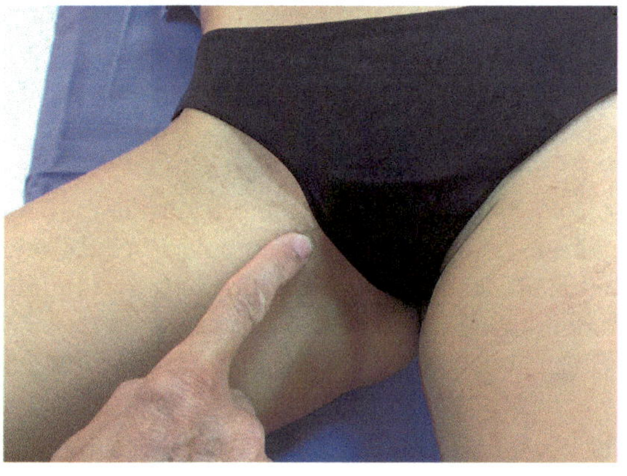

Fig. 14.1 Adductor longus

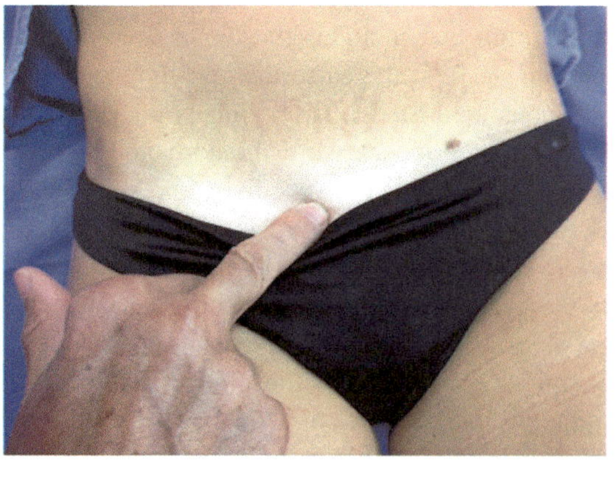

Fig. 14.2 Rectus abdominis

14.2 Clinical Assessment and Diagnosis

Because of the several anatomical structures in the groin area, the diagnosis is often difficult, and attention to differential diagnosis must be paid [8].

The physical examination should rule out a hypothetic inguinal or abdominal hernia, as well as the femoroacetabular impingement (FAI) with a concomitant labral tear or a pubic symphysis affection [12].

The most common clinical presentation is groin or lower abdominal pain, often associated with a pain irradiation to the medial aspect of the thigh, to the abdomen, or even to the perianal area [9].

Usually unilateral, pain starts after physical activity and moves progressively to the other side, often compromising everyday activities and social life. In advanced stages, even actions like sneezing, coughing, defecating, or sexual activity cause pain [13].

For a proper evaluation and a correct diagnosis, several factors must be taken into account.

Clinical evaluation is important to determine whether the cause is articular, musculotendinous, hernia related, or of combined origin [14].

Observation of the patient is chiefly important in order to evaluate the mobility of the hips on all planes, the plantar support, the hindfoot and forefoot structure, as well as the posture, the rachis mobility, the spinal curvatures, and the modified rotation of the pelvis, if present. In fact it is well demonstrated that the adductor tendinopathy is commonly correlated with lumbar hyperlordosis and anteversion of the pelvis [9]. Furthermore, the clinician should observe the inguinal canal, the symmetry of posterior superior iliac spines with the patient in supine position, the knee axis, and the patellar orientation.

The palpation of some specific point helps to identify the most painful areas. The most frequently finding for adductor tendinopathy is tenderness along the tendon during passive abduction and resisted hip adduction in extension [15] (Fig. 14.3). Also rectoabdominal and iliopsoas tendon insertions are considerable typical painful points in this tendinopathy [5] (Fig. 14.2). Pain may be also evocated during the adduction or contraction of the abdominal muscles (Fig. 14.4), iliopsoas (Fig. 14.5), and rectus

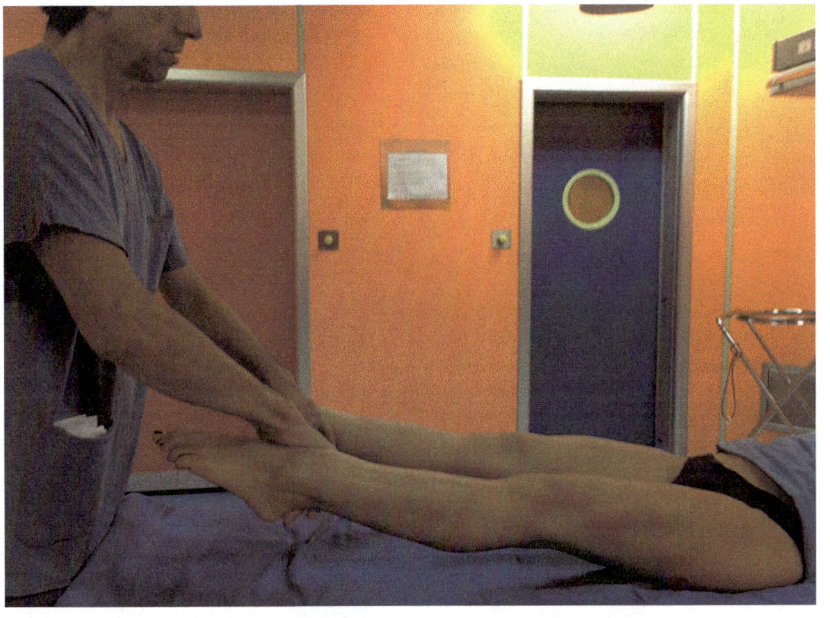

Fig. 14.3 Adductor muscle and tendon strain associated with medial thigh and anterior groin pain during resisted adduction and passive abduction

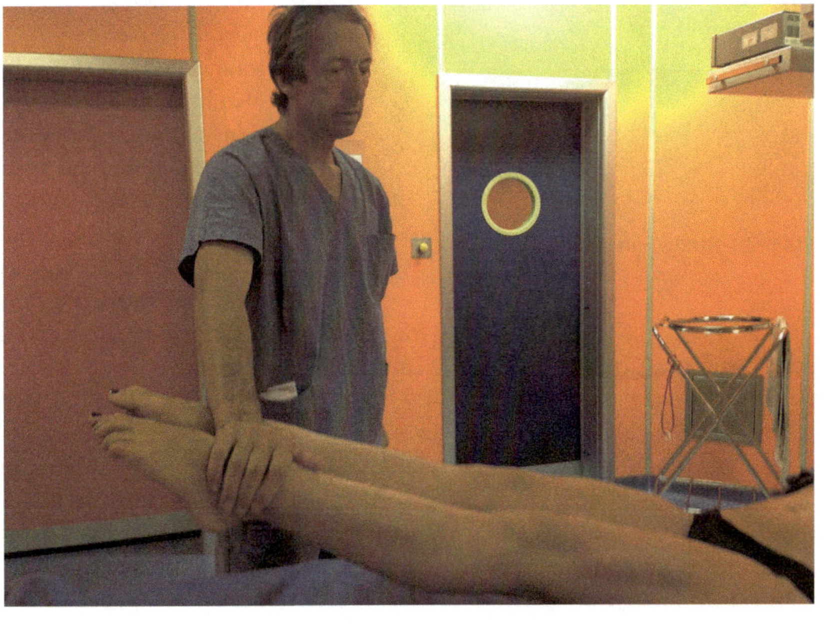

Fig. 14.4 Rectus abdominis. Pain with resisted trunk flexion and localized tenderness to palpation

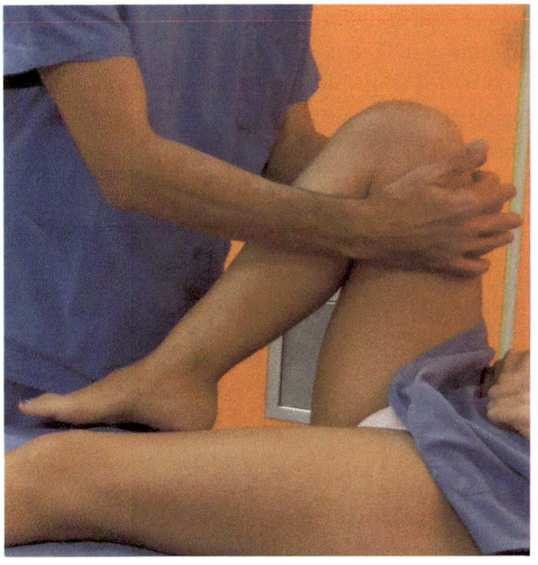

Fig. 14.5 Iliopsoas strain: pain with hip hyperextension or resisted hip flexion at 90°

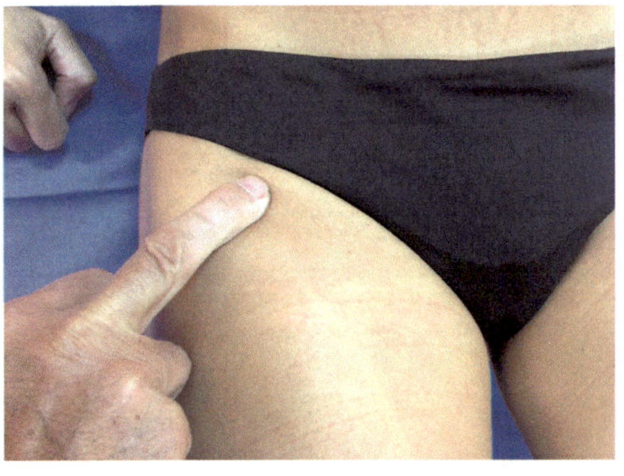

Fig. 14.6 Rectus femoris

femoris (Figs. 14.6 and 14.7), while the clinician exerts counter resistance during passive stretching of the adductors and iliopsoas muscles [9].

Specific clinical tests are for the evaluation of the articular mobility and to identify more accurately the painful points: through these tests it is possible to prove the shortening of the anterior chain, the posterior chain, and the sacroiliac joint [9]. Similarly, the peripheral neurological examination may be performed using the Lasegue test, the Wassermann test, the patellar and Achilles tendon reflexes, and the cutaneous sensibility. As regards the adductor muscles, it is mostly common to evaluate the isometric contraction against proximal or distal resistance, with flexed or extended legs, for the assessment of the abdominal muscles and the eccentric contraction of both the rectus abdominis and the oblique muscles.

The instrumental confirmation of the diagnosis consists of conventional radiographs, ultrasound, and magnetic resonance.

X-ray assessment is used to confirm the diagnosis and to exclude bone or coxofemoral joint disorders [14]. Plain radiographs show the hips and pelvis symmetry and observe the tendon insertional areas and the bone structure. The anteroposterior weight-bearing view of the pelvis is particularly useful. It is not uncommon to locate hip osteoarthritis and modified limiting bone in the presence of groin pain [9]. The flamingo stress views are also used to evaluate pelvic stability [16].

Ultrasound is the first-line imaging modality to assess musculotendinous structures, insertional tendon areas, ligaments, and all the soft tissues. It is useful to distinguish between an acute trauma and a chronic overload injury, and it may also individuate an inguinal hernia or any alteration of the posterior wall of the inguinal canal [17]. Longitudinal and axial images are then compared to the contralateral side.

Magnetic resonance imaging (MRI) is used to investigate some morphological details, because of the elevated contrast resolution images. It is usually helpful to individuate bone marrow edema, insertional tendinopathy, and symphysis capsular disruption [9] (Fig. 14.8). Central disc protrusion is relatively common among soccer players [18].

MRI is generally preferred over computed tomography (CT) in case of acute lesions; conversely, CT images are commonly used for calcific or insertional chronic tendinopathies because of its detailed evaluation [19].

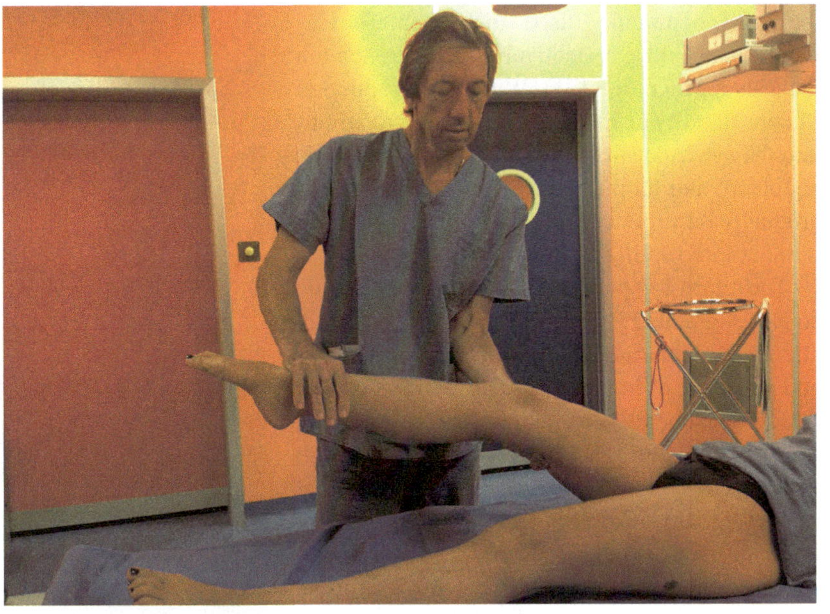

Fig. 14.7 Rectus femoris strain: pain with resisted hip flexion and knee extension

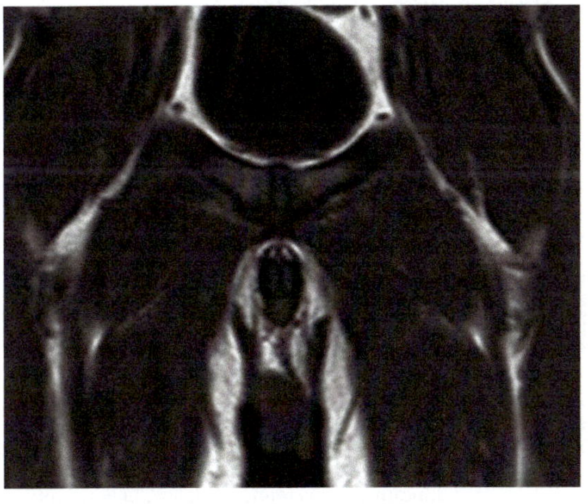

Fig. 14.8 MRI imaging: bilateral adductor tendinopathy associated with osteitis pubis

14.3 Management

The management of adductor and abdominal tendinopathies consists of multimodal measures, such as pharmacological therapy, physical rehabilitation, and instrumental procedures.

The rehabilitation period may be distinguished in acute phase, subacute phase, and return to sport. During the whole healing time, a neuromuscular taping may be applied in order to promote muscular relaxation and to protect the musculotendinous units from any overstretching [20].

In the acute phase, the most important goal is pain reduction, mainly through muscular relaxation. This may be obtained using local injections of NSAIDs, corticosteroids, supplements for muscles and tendons, or platelet-derived growth factor (PDGF) [9].

Rehabilitation starts with postural balance measures, such as global and site-specific stretching, decontracting massotherapy, mechanical and proprioceptive recovery, and global postural exercises. During this first stage, isometric strengthening of the abdominal and adductor muscles is advised.

In the subacute phase, concentric and eccentric exercises may be started to promote muscle strengthening, always paying attention to the cardiovascular system. Massotherapy is also useful to stimulate microcirculation and to reduce inflammation [9]. Heat therapy, extracorporeal shock wave therapy, and laser therapy are the most commonly used instrumental treatments, since they effectively promote tendon enthesis regeneration [21]. With the introduction of core stability exercises, it is possible to stimulate the synergic strengthening of the adductors, abdomen, and lumbar muscles [22]. Running is gradually reintroduced, starting on the treadmill, without interrupting any instrumental therapy.

The return-to-sport phase reintroduces aerobic activity, gradually increasing running speed, alternated with anaerobic training, stretching, and repeated exercises [9]. The neuromuscular system must be fully recovered, according to the sport-specific functional requests, also through tendon overloading exercises. It is fundamental to prevent any recurrent lesion by performing training matches and preventive postural and eccentric strengthening and plyometric exercises, in order to guarantee a good stability of the posterior chain and an ideal balance between agonist and antagonist muscle groups [23].

It is possible that conservative measures do not guarantee a safe and effective return to sport. In this case, after at least 3 months of failed conservative management, surgery may be considered. In particular, the surgical technique for the adductor tendinopathy consists of a release of the adductor longus, through a detensioning open or percutaneous tenotomy. It is important to release partially the tendon; the anterior fibers are sectioned, leaving the deeper muscular body intact (Figs. 14.9, 14.10, and 14.11). After the surgical procedure, a minimum loss of the adductor longus strength can occur, although other muscles such as the adductor brevis, adductor magnus, and pectineus easily compensate this reduction [24].

In case of a concomitant inguinal hernia, several authors recommend the surgical placement of a polypropylene or biologic mesh over the posterior wall of the inguinal canal, in order to strengthen the area, relieving pressure and improving the inflammatory response [25, 26], and mechanically tighten the abdominal sheet.

Some authors have recently proposed the resection of degenerative soft tissue and the subsequent reattachment with suture anchors [27], possibly associated with arthroscopic debridement of the pubic symphysis. Other authors prefer to perform a bilateral adductor tenotomy even if the affection is unilateral as a preventive measure [2].

Generally, the patient may return to the preinjury sport activity after 3 months [9]. After following adductor stretching programs, closed chain adductor strengthening exercises, straight-

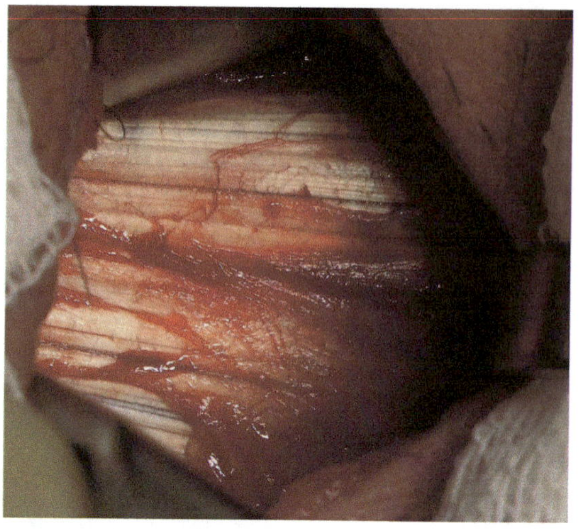

Fig. 14.9 Thickened adductor longus tendon

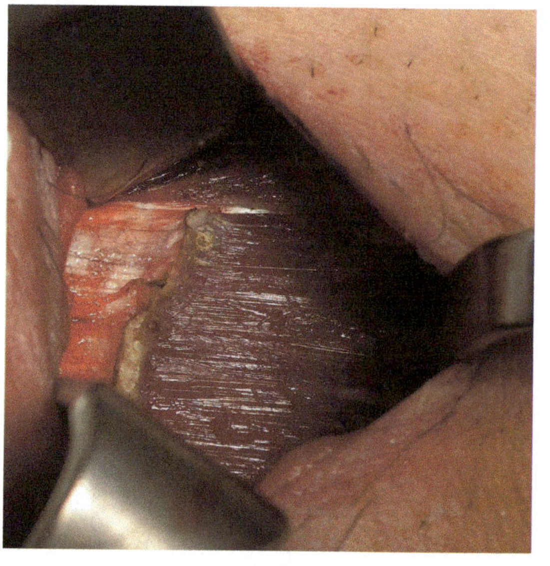

Fig. 14.10 Adductor longus tendon release. Deeper muscular body left intact

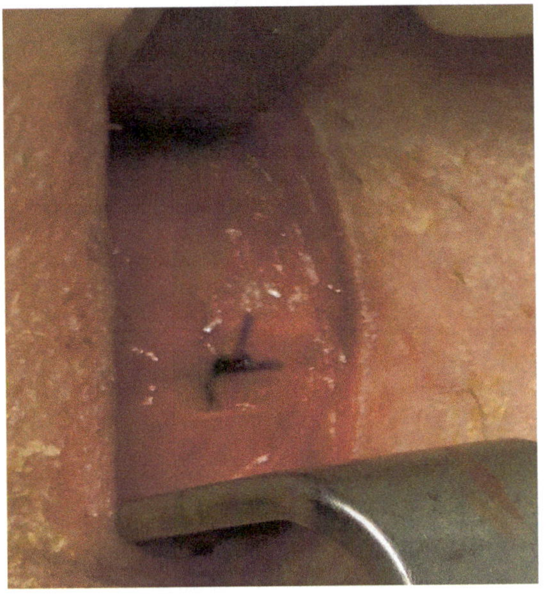

Fig. 14.11 Accurate peritenon suture

line jogging, and then open chain exercises may be started once the closed chain movements do not cause any pain or discomfort. After a mean time of 3 weeks, the pace of running may be increased, introducing also side-to-side movements [28].

14.4 Prevention Programs

It has been well documented that increasing adductor muscle strength reduces the risks of recurrent lesions. Tyler et al. [10] reported a specific prevention program for ice hockey players, divided into three distinct stages, in order to obtain an adduction-to-abduction strength ratio of at least 80% and to ensure the proper balance between agonist and antagonist muscle groups.

The "warm-up" phase consists of biking, squats, side lunges, kneeling pelvic tilts, and adductor strengthening exercises. The strengthening program requires ball squeezes, concentric adduction against resistance, and several other adduction exercises using different machines.

The sport-specific training, finally, consists of crossover pulls, slide skating, and standing resisted stride lengths. Obviously every sport requires a different specific training.

Conclusions

The tendinopathy of the adductors and the rectus abdominis is a common cause of groin pain in athletes. If preventive measures and conservative treatment fail, an open or percutaneous approach may be effective and allows return to sport activities.

References

1. Estwanik JJ, Sloane B, Rosemberg MA. Groin strain and other possible causes of groin pain. Phys Sportsmed. 1988;18:59–65.
2. Maffulli N, Loppini M, Longo UG, et al. Bilateral mini-invasive adductor tenotomy for the management of chronic unilateral adductor longus tendinopathy in athletes. Am J Sports Med. 2012;40(8):1880–6.
3. Robinson P, Barron DA, Parsons W, et al. Adductor-related groin pain in athletes: correlation of MR imaging with clinical findings. Skeletal Radiol. 2004;33(8):451–7.
4. Orso CA, Calabrese BF, D'Onofrio S, et al. The rectus-adductor syndrome. The nosographic picture and treatment criteria. Ann Ital Chir. 1990;61:657–9.
5. Maffulli N, Giai Via A, Oliva F. Groin pain—insertional adductors and rectus abdominis tendinopathy. In: Volpi P, editor. Football traumatology: new trends. 2nd ed. Cham (ZG): Springer; 2015. p. 305–6.
6. Weir A, Brukner P, Delahunt E, et al. Doha agreement meeting on terminology and definitions in groin pain in athletes. Br J Sports Med. 2015;49(12):768–74.
7. Orchard J, Read JW, Verrall GM, et al. Pathophysiology of chronic groin pain in the athlete. ISMJ. 2000;1(1):1–16.
8. Morelli V, Smith V. Groin injuries in athletes. Am Fam Physician. 2001;64:1405–14.
9. Valent A, Frizziero A, Bressan S, et al. Insertional tendinopathy of the adductors and rectus abdominis in athletes: a review. Muscles Ligaments Tendons J. 2012;2(2):142–8.
10. Tyler TF, Fukunaga T, Gellert J. Rehabilitation of soft tissue injuries of the hip and pelvis. Int J Sports Phys Ther. 2014;9(6):785–97.
11. Riley G. The pathogenesis of tendinopathy. A molecular perspective. Rheumatology. 2004;43:131–42.
12. Gill TJ, Carroll KM, Makani A, et al. Surgical technique for treatment of recalcitrant adductor longus tendinopathy. Arthrosc Tech. 2014;3(2):e293–7.

13. Robertson BA, Barker PJ, Fahrer M, et al. The anatomy of the pubic region revisited: implications for the pathogenesis and clinical management of chronic groin pain in athletes. Sports Med. 2009;39:225–34.
14. Pesquer L, Reboul G, Silvestre A, et al. Imaging of adductor-related groin pain. Diagn Interv Imaging. 2015;96(9):861–9.
15. Serner A, Tol JL, Jomaah N, et al. Diagnosis of acute groin injuries: a prospective study of 100 athletes. Am J Sports Med. 2015;43(8):1857–64.
16. Garvey JF, Read JW, Turner A. Sportsman hernia: what can we do? Hernia. 2010;14:17–25.
17. Bradley M, Morgan J, Pentlow B, et al. The positive predictive value of diagnostic ultrasound for occult herniae. Ann R Coll Surg Engl. 2006;88:165–7.
18. Branci S, Thorborg K, Bech BH, et al. MRI findings in soccer players with long-standing adductor-related groin pain and asymptomatic controls. Br J Sports Med. 2015;49:57–64.
19. Benazzo F, Mosconi M, Zanon G, et al. Groin pain. J Sports Traumatol. 1999;21(1):30–40.
20. Morrissey D, Graham J, Screen H. Coronal plane hip muscle activation in football code athletes with chronic adductor groin strain injury during standing hip flexion. Man Ther. 2012;17:145–9.
21. Weir A, Jansen J, van Keulen J, et al. Short and mid-term results of a comprehensive treatment program for longstanding adductor-related groin pain in athletes: a case series. Phys Ther Sport. 2010;11(3):99–103.
22. McGill SM. Low back stability: from formal description to issues for performance and rehabilitation. Exerc Sport Sci Rev. 2001;29:26–31.
23. Weir A, Jansen JA, van de Port IG, et al. Manual or exercise therapy for long-standing adductor-related groin pain: a randomised controlled clinical trial. Man Ther. 2011;16:148–54.
24. Atkinson HD, Johal P, Falworth MS, et al. Adductor tenotomy: its role in the management of sports-related chronic groin pain. Arch Orthop Trauma Surg. 2010;130(8):965–70.
25. Richardson WS, Jones DG, Winters JC, et al. The treatment of inguinal pain. Ochsner J. 2009; 9(1):11–3.
26. Edelman DS, Selesnick H. "Sport" hernia: treatment with biologic mesh (Surgisis): a preliminary study. Surg Endosc. 2006;20:971–3.
27. Hopp SJ, Culemann U, Kelm J, et al. Osteitis pubis and adductor tendinopathy in athletes: a novel arthroscopic pubic symphysis curettage and adductor reattachment. Arch Orthop Trauma Surg. 2013; 133(7):1003–9.
28. Schilders E, Dimitrakopoulou A, Cooke M, et al. Effectiveness of a selective partial adductor release for chronic adductor-related groin pain in professional athletes. Am J Sports Med. 2013;41(3):603–7.

15 Iliopsoas Tendinopathy

Gianni Nanni

15.1 Introduction

There is no epidemiological data in the literature on the actual incidence of the iliopsoas tendinopathy in both the sport and non-sportive population.

However, it is more commonly found in young people with prevalence in females.

The incidence of this pathology is more prominent in athletics: football/soccer; artistic gymnastics; canoeing; running, especially uphill; dance and in all those activities that require repetitive movements of flexion or extra-rotation of the hip joint [1–3].

15.2 Anatomy

The iliopsoas muscle is made up of two distinct parts: the psoas major muscle and the iliac muscle which unite and insert on the lesser trochanter (Fig. 15.1).

The psoas major muscle is fusiform and originates through a series of fibrous arches, from the lateral aspect of the bodies of the last thoracic vertebra and the first four lumbar vertebrae and from the interposed intervertebral discs and

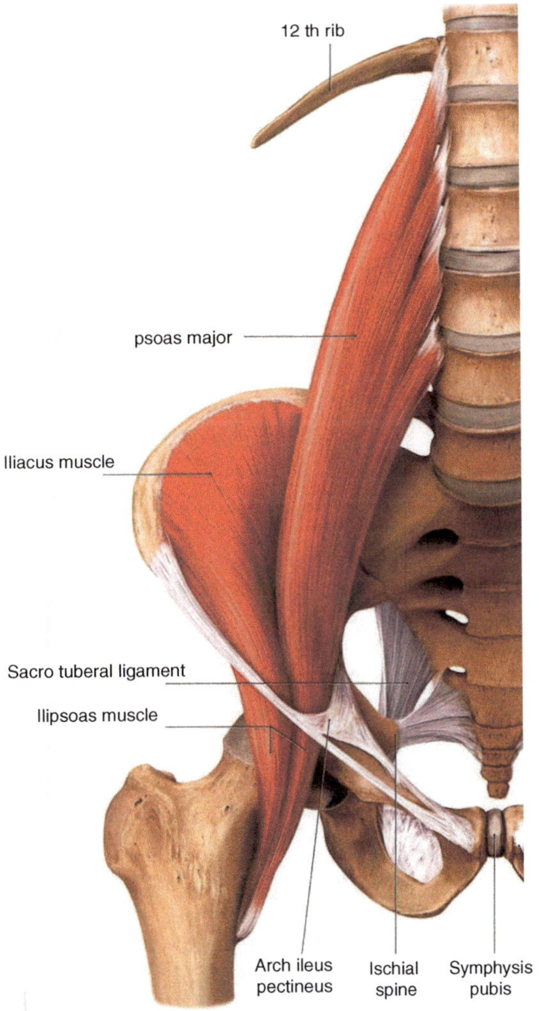

Fig. 15.1 Iliopsoas anatomy

G. Nanni, M.D.
Isokinetic Medical Group, FIFA Medical Centre of Excellence, Bologna, Italy
e-mail: g.nanni@isokinetic.com

R. Zini et al. (eds.), *Groin Pain Syndrome*, DOI 10.1007/978-3-319-41624-3_15

the base of the transverse processes of the first four lumbar vertebrae. It continues caudally and laterally through the lumbar and iliac region. It exits the pelvis passing under the inguinal ligament, between the anterior inferior iliac spine and the iliopectineal eminence. It then passes anterior to the hip joint and inserts into the lesser trochanter, after a characteristic rotation in its course, so that its ventral surface becomes medial [4, 5].

The iliopectineal bursa, the largest in the human body, is interposed between its robust distal tendon and the hip joint fibrous capsule [1, 5] (Fig. 15.2).

The fan-shaped iliacus muscle originates from the inner lip of the iliac crest, the two anterior iliac spines, and from the notch interposed between them, the upper 2/3 of the iliac fossa, the iliolumbar ligament, and the lateral portion of the sacral wing.

From this extensive origin line, the muscular fibers converge inferiorly and terminate by partially merging with those of the psoas major muscle, while the most lateral muscle fibers insert on the lesser trochanter without joining with the main tendon.

The iliopsoas muscle is innervated by branches of the lumbar plexus and then the femoral nerve (L1–L4) [4].

15.3 Symptoms and Clinical Examination

Iliopsoas tendinopathy occurs most frequently as a result of overuse and, as we have already noted, more often in those sports that require repetitive hip flexion and external rotation movements [1].

More rarely, the pain starts following a single eccentric iliopsoas contraction. In this case the

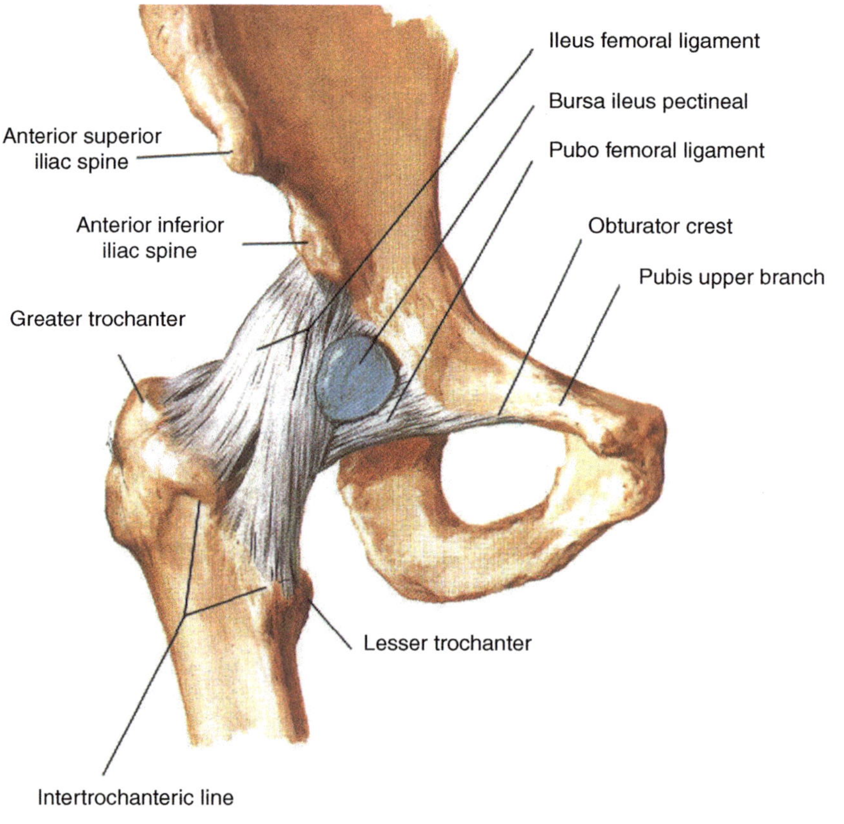

Fig. 15.2 Iliopsoas bursa

pain is most frequently caused by a myotendinous junction lesion of the iliopsoas itself.

This chapter includes: the insertional tendinopathy with inflammatory characteristics affecting the distal bone-tendon junction at the level of the lesser trochanter; the peritendinitis with possible and frequent involvement of the iliopectineal mucous bursa that is anterior to the hip joint capsule and lies between the iliopsoas tendon and the bone margin of the iliopubic branch; the tendinosis, actual fibrillar tissue degeneration of the tendon; the peritendinitis and tendinosis: the partial or total (very rare) lesion of the tendon itself.

The iliopsoas muscle tendon unit may also be affected by true injuries to the muscle or to the myotendinous junction by direct or, more frequently, by indirect trauma. If not properly rehabilitated, reaching full extensibility and strength recovery, it could then cause tendinopathies of the iliopsoas tendon or lead to "internal snapping hip."

Internal snapping hip is the patient's perception of a "clicking" sound in the groin also audible by the operator, which can be more or less painful depending on the degree of inflammation of the iliopsoas tendon [1, 6–8].

The snapping occurs when moving the hip joint from a position of flexion, extra-rotation, and hip abduction to a position of extension, intra-rotation, and adduction, and it can be perceived when rising from sitting to standing, especially with the hip at an acute angle (sitting low) [9].

The snapping is due to the rubbing of the iliopsoas tendon on the iliopectineal eminence or to the movement of the psoas tendon onto the iliac tendon.

The Thomas test can be positive [9, 10] (Fig. 15.3).

This condition must be differentiated from the hip snap due to intra-articular pathologies such as injuries of the acetabular labrum or as cartilaginous loose bodies.

The snapping hip can also be due to an "external snapping hip" caused by the tensor fasciae latae muscle tendon passage on the femoral greater trochanter, diagnosed by the Ober test [1, 9, 10].

In iliopsoas tendinopathies, the pain is located in the groin and below the pubic area with possible radiation to the medial thigh.

Most often the pain occurs subtly and does not prevent any sport activity.

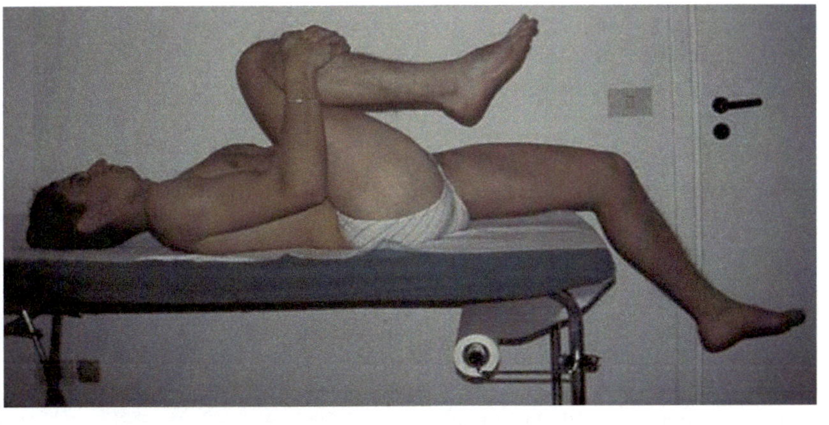

Fig. 15.3 Thomas test. The Thomas test is a physical examination test used to highlight short hip flexor and iliopsoas tendinopathy. The patient lies supine on the examination table and holds the uninvolved knee to the chest while allowing the involved extremity to lie flat. Holding the knee to the chest flattens out the lumbar lordosis and stabilizes the pelvis. If the iliopsoas muscle is shortened, the lower extremity on the involved side will be unable to fully extend the hip, which will remain raised from the couch

Subsequently a progressive increase in the intensity of pain is reported during sporting activity with pain even at rest. This establishes compensatory postural mechanisms which over time determine changes in pain characteristics giving rise to pain in other locations of the pelvis as well (contralateral, suprapubic, etc.).

During hip movement, pain can be coupled by a noise (internal snapping hip) [1, 8, 9].

It can be difficult to understand whether a groin pain comes from iliopsoas or adductor longus tendinopathy or another tendon which inserts onto the pubis.

The contraction tests against manual resistance can be helpful.

A positive iliopsoas contraction test elicits pain with flexion at the externally rotated hip, and the knee fully extended is positive, while adductor and abdominal contraction against resistance is less painful and shows less functional impairment [11, 12].

Furthermore, the Thomas test is almost always positive on the side of tendinopathy [1, 10].

The Liedloff test is positive with the patient seated and the knee extended as the patient is asked to lift the heel off the ground causing pain and functional impairment on the affected side.

A characteristic snap is classically caused by passing from a flexed hip position, abducted and externally rotated, to an extended one, adduced and internally rotated.

When evaluating a patient with iliopsoas tendinopathy, the back and pelvis mechanics should be checked.

In fact the psoas major muscle is the only one that connects the spine with the lower limb. Besides being a flexor and a femur rotator, it is also a spine flexor and an important postural muscle activated during many movements and daily activities [13].

It is in fact a muscle with a prevalence of type 1 fibers (slow twitch) [1].

In the trunk flexion movement (Piédallu test), patients with reduced iliopsoas extensibility tend to have an anterior ileum with an increased excursion of the ipsilateral PSIS (posterior superior iliac spine) [14] and a tendency of walking with a slightly flexed knee.

Tests for core stability are also important as they can highlight iliopsoas changes in strength, even unilaterally [15].

These patients can both present with the typical inguinal symptoms and ipsilateral back pain [11].

15.4 Differential Diagnosis

The differential diagnosis should be made with hip intra-articular orthopedic pathologies such as FAI (femoral acetabular impingement) [1, 9, 16], acetabular labrum tears, femoral head edema, osteochondral detachments, or systemic conditions such as rheumatic or nonrheumatic osteoarthritis [17].

Differential diagnoses of the iliopsoas tendinopathies include insertional tendinopathies; proximal myotendinous lesions of the adductor muscle, the rectus femoris, and the sartorius; distal myotendinous lesions of the abdominal muscle; and obturator muscolotendinous pathologies [1, 16].

In the period of age growth, the detachment of the lesser trochanter following a violent iliopsoas contraction with hip flexion resistance can occur, although, rarely, mostly during sport activities (Fig. 15.4).

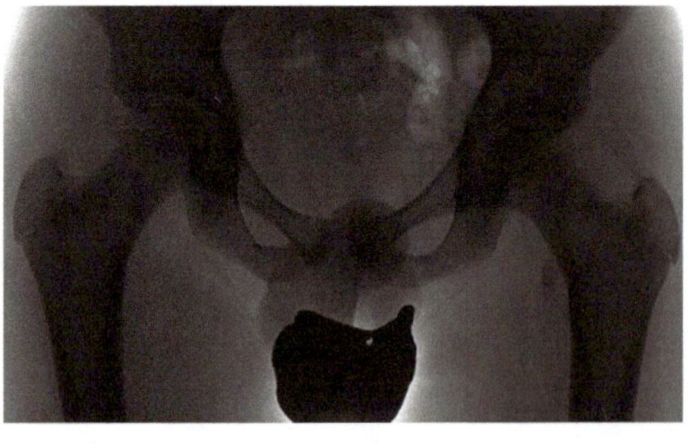

Fig. 15.4 X-ray: left femoral lesser trochanter detachment

15.5 Imaging

X-rays can be useful to differentiate a groin pain due to FAI (anteroposterior pelvic X-rays and frog lateral view with the hip abducted to 45°) [9] or document an avulsion fracture (e.g., of the lesser trochanter).

Ultrasound: it is useful in imaging iliopsoas muscle-tendon injuries of the extra pelvic tract. It can show a thickening of the iliopsoas tendon and/or an associated bursitis, with both an increase of the liquid component within the bursa and its volume.

A dynamic ultrasound can be nullifying in documenting an internal snapping hip [7, 18, 19].

It can be useful in the differential diagnosis with other tendon muscle pathologies, which for anatomical reasons can cause groin pain.

It is a difficult area to study with ultrasound, and being operator dependent, it requires a certain amount of experience.

MRI: it is the gold standard in the psoas tendinopathy evaluation and is particularly useful in the differential diagnosis with other pathologies that cause groin pain [20, 21] (Fig. 15.5).

It is essential in documenting an iliopsoas injury affecting the intrapelvic portion before passing under the inguinal ligament where an ultrasound may not provide diagnostic images [22] (Fig. 15.6).

In a muscle tendon injury with edema and inflammation, the T2-weighted images show an increase in the intensity of the signal. In the case of blood extravasation, both the T1- and T2-weighted images show an elevated intensity of the signal.

In evaluating peritendinitis, the liquid pooling in the peritendineum tissue is detected in the T2 or STIR images as an area of high signal intensity surrounding a normal tendon.

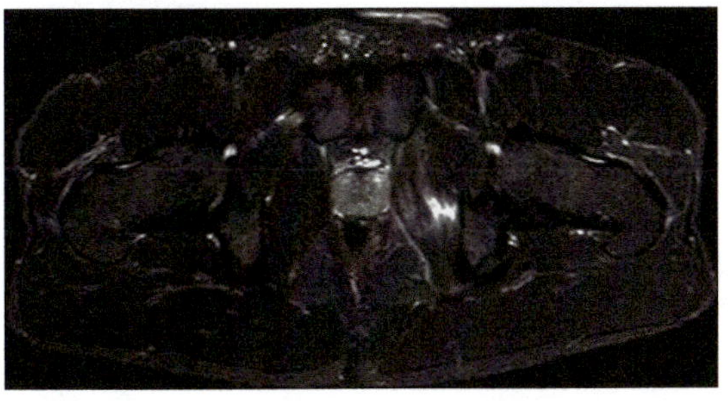

Fig. 15.5 MRI: obturator internus muscle injury in the left side

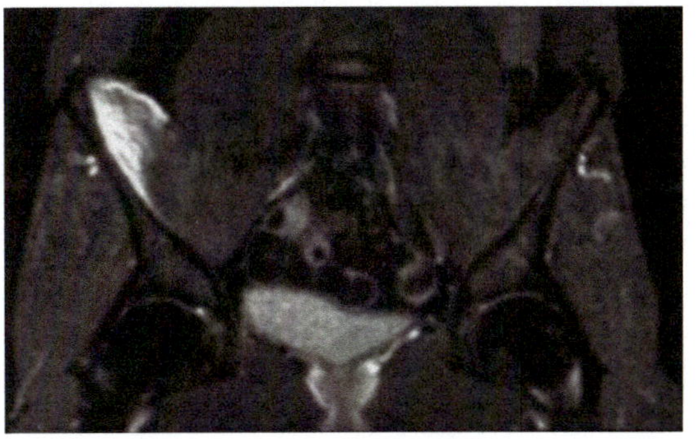

Fig. 15.6 MRI: right iliac muscle contusion in a professional soccer player

Tendinosis, in T1-weighted images, shows an increase of signal intensity within the tendon associated with myxoid degeneration or angiofibroblastic proliferation. The T2 images may have a normal or less intense signal than those in the T1-weighted ones [21].

15.6 Therapy

Most patients benefit from conservative treatment. Surgery is required only in rare cases where pain is not resolved and they fail to resume sport activities or when suffering from a painful and debilitating internal snapping hip.

Surgery is rare in iliopsoas tendinopathies. Surgery consists of a tendon release, mostly endoscopically [23, 24].

Rehabilitation treatment is divided into five steps:

- Step 1: pain, swelling, and inflammation reduction
- Step 2: range of motion and extensibility recovery
- Step 3: strength and resistance recovery
- Step 4: coordination and proprioception recovery
- Step 5: specific technical movements and athletic sport-specific parameters recovery

In step 1 the aim is to reduce pain and inflammation. It is important to let the patient rest especially if it concerns an athlete. Provide crutches if the pain alters proper walking pattern. Use

NSAIDs for the first 5–7 days in those few patients with very intense pain, and apply cryotherapy.

The second step involves exercises that restore extensibility and full range of motion of the hip and the lumbosacral spine. Iliopsoas tendinopathies often show a reduction of extensibility and an anterior ipsilateral ilium. Stretching exercises of the iliopsoas, rectus femoris, and tensor fascia latae are therefore important, as well as including chiropractic treatment of the pelvis and lumbosacral spine. If there are no muscle or myotendinous junction injuries, iliopsoas myofascial massage helps to promote relaxation of the muscle-tendon units, frequently the seat of trigger points [1].

Stretching exercises should be performed under the threshold of pain (without pain) and must not be done after cryotherapy application. In this phase massotherapy is very important and must precede the execution of the exercises. Strength recovery of both the iliopsoas and core muscles is very important. Core stability exercises must precede those for iliopsoas strength recovery, and an optimum selective abdominal muscle control should be obtained before undertaking those for the iliopsoas [15, 25].

It is imperative to perform exercises without pain and progress very gradually. Exercises to improve iliopsoas strength are based on hip flexion. The exercise starts in the supine position with bigger knee and hip flexion angles at first, until being able to perform the exercise with the knee fully extended and the hip in line with the body, with the leg flat on the couch. The progression continues in a sitting position with toning exercises, sitting on an unstable surface (e.g., Bobath ball), then standing against the wall, standing without support, then standing on a foam mat, and finally using a ballasted anklet or a rubber band (Fig. 15.7). It is important to observe the patient while performing the hip flexor exercises in order to execute them with an excellent pelvis and spine control [15]. When the progression of the exercises with a good pelvic and trunk movement control has been completed without problems, the athlete patient will have to start recovering technical sport-specific gestures, again with proper progression, before returning to free training [22]. Finally, the iliopsoas bursitis may require an infiltrative-guided corticosteroid treatment.

Fig. 15.7 Hip flexion progression for iliopsoas tendinopathy

References

1. Tyler TF, Fukunaga T, Gellert J. Rehabilitation of soft tissue injuries of the hip and pelvis. Int J Sports Phys Ther. 2014;9(6):785–97.
2. Rose DJ, Montalbano G, Rosen J, Peters WG, Liederbach M. Iliopsoas syndrome in dancers. Med Sci Sports Exer. 1998;30(5 Suppl):288.
3. Brunot S, Dubeau S, Laumonier H, Creusé A, Delmeule T, Reboul G, Das Neves D, Bouin H. Acute inguinal pain associated with iliopectineal bursitis in four professional soccer players. Diagn Interv Imaging. 2013;94(1):91–4.
4. Manzoli FA, et al. Anatomia Umana. Edi-Ermes. 2004;1:325–6.
5. Tatu L, Parratte B, Vuillier F, Diop M, Monnier G. Descriptive anatomy of the femoral portion of the iliopsoas muscle. Anatomical basis of anterior snapping of the hip. Surg Radiol Anat. 2001;23(6):371–4.
6. Jacobs M, Young R. Snapping hip phenomenon among dancers. Am Correct Ther J. 1978;32(3):92–8.
7. Janzen DL, Partridge E, Logan PM, Connell DG, Duncan CP. The snapping hip: clinical and imaging findings in transient subluxation of the iliopsoas tendon. Can Assoc Radiol J. 1996;47(3):202–8.
8. Winston P, Awan R, Cassidy JD, Bleakney RK. Clinical examination and ultrasound of self-reported snapping hip syndrome in elite ballet dancers. Am J Sports Med. 2007;35(1):118–26.
9. Zini R. Hip arthroscopy. Argalìa Editore Urbino. 2010;Chapter 2. 30–1.
10. Malanga GA, Nadler SF. Musculoskeletal physical examination: an evidence-based approach. Philadelphia: Elsevier Mosby; 2006. p. 251–79.
11. Laible C, Swanson DJ, Rose D. Iliopsoas syndrome in dancers. Orthop J Sports Med. 2013;1(3):1–6.
12. Kendall HO, Kendall FP, Wadsworth GE. I muscoli. Esame e studio funzionale. PICCIN. 1985;Chapter 5. 176–189.
13. Andersson E, Oddsson L, Grundström TA. The role of the psoas and iliacus muscles for stability and movement of the lumbar spine, pelvis and hip. Scand J Med Sci Sports. 1995;5(1):10–6.
14. Maigne R. Medicina Manuale. UTET. 1994; Appendix. Chapter 6. p. 499.
15. Willson JD, Dougherty CP, Ireland ML, Davis IM. Core stability and its relationship to lower extremity function and injury. J Am Acad Orthop Surg. 2005;13(5):316–25.
16. Anderson K, Strickland SM, Warren R. Hip and groin injuries in athletes. Am J Sports Med. 2001;29(4):521–33.
17. Generini S, Matucci-Cerinic M. Iliopsoas bursitis in rheumatoid arthritis. Clin Exp Rheumatol. 1993;11(5):549–51.
18. Deslandes M, Guillin R, Cardinal E, Hobden R, Bureau NJ. The snapping iliopsoas tendon: new mechanisms using dynamic sonography. AJR Am J Roentgenol. 2008;190(3):576–81.
19. Fredberg U, Hansen LB. Ultrasound in the diagnosis and treatment of iliopsoas tendinitis: a case report. Scand J Med Sci Sports. 1995;5(6):369–70.
20. De Paulis F, Cacchio A, Michelini O, Damiani A, Saggini R. Sports injuries in the pelvis and hip: diagnostic imaging. Eur J Radiol. 1998;27(Suppl 1):S49–59.
21. Shin AY, Morin WD, Gorman JD, Jones SB, Lapinsky AS. The superiority of magnetic resonance imaging in differentiating the cause of hip pain in endurance athletes. Am J Sports Med. 1996;24(2):168–76.
22. Maffulli N, et al. Linee guida I.S.Mu.L.T. per il trattamento delle lesioni muscolari. Exer Med. 2014;68–69, 93–105.
23. Ilizaliturri Jr VM, Camacho-Galindo J. Endoscopic treatment of snapping hips, iliotibial band, and iliopsoas tendon. Sports Med Arthrosc. 2010;18(2):120–7.
24. Anderson SA, Keene JS. Results of arthroscopic iliopsoas tendon release in competitive and recreational athletes. Am J Sports Med. 2008;36(12):2363–71.
25. Heiderscheit BC, Sherry MA, Silder A, Chumanov ES, Thelen DG. Hamstring strain injuries: recommendations for diagnosis, rehabilitation, and injury prevention. J Orthop Sports Phys Ther. 2010;40(2):67–81.

Part 5

Bone Pathology

16 Osteitis Pubis

Gianluca Melegati and Sara Elli

Osteitis pubis is a painful inflammatory disorder that affects the pubic symphysis and the insertions of the surrounding adductor and abdominal muscles. This well-defined clinical entity is one of the main causes of rectus-adductor-symphysis syndrome, commonly known as "groin pain" [1].

The typical presentation of the disorder is progressive pain with insidious onset in the pubic adductor region, often radiating along the adductor muscles and into the suprapubic region. Pain is exacerbated by dynamic stresses such as kicking, side-to-side movements, accelerations and decelerations, leaps and pivoting on one leg. However, the most effective clinical test for determining the disorder is the onset of pain when standing on one leg, while dressing.

Some authors consider the disorder to be self-limiting, with the presence of erosions affecting one or both edges of the joint with resulting repair processes [2].

G. Melegati (✉)
Department of Rehabilitation Medicine,
Galeazzi Orthopaedic Institute—IRCCS, Milan, Italy
e-mail: gianluca@melegati.it

S. Elli
School of Specialisation in Physical and Rehabilitation Medicine, University of Milan, Milan, Italy

16.1 Epidemiology

Today osteitis pubis is considered the most common inflammatory disease of the pubic symphysis. Although it can theoretically occur at any age, it is very rare in the paediatric population and typically occurs in males during the third and fourth decades of life. Its incidence in the athletic population has not been widely investigated; according to some authors, it ranges from 0.5 to 6.4% [4] and is higher among footballers; basketball, hockey and rugby players; and long-distance and middle-distance runners.

Orchard [3] gave an indirect estimate of the incidence of osteitis pubis in a population of Australian footballers studied over a period of 4 years. In his epidemiological study, osteitis pubis was included in a nonspecific category indicated as "other groin/hip/thigh injuries", with an incidence of 0.2%.

One epidemiological study conducted on 189 athletes with groin pain demonstrated that osteitis pubis was the first cause of pain in 14% of athletes [4].

A magnetic resonance imaging study of 97 athletes affected by groin pain found that 9.3% were affected by isolated osteitis pubis, whereas in 42.3% osteitis pubis was associated with concurrent microtear of the adductor attachment [5].

This demonstrates that osteitis pubis is a common clinical entity in the athletic population, with a male predominance that is unlikely to be justified by a mere proneness related to the male

R. Zini et al. (eds.), *Groin Pain Syndrome*, DOI 10.1007/978-3-319-41624-3_16

sex but more probably results from a gender difference in sports participation which continues to be evident.

16.2 Aetiopathogenesis

Osteitis pubis was first described in 1924 by a urologist, Beer [6], who reported on five patients who had developed a form of osteitis pubis after undergoing gynaecological and urological surgery, and the disorder is still known to be a possible complication of invasive pelvic procedures such as suprapubic herniorrhaphy, prostatectomy and cystectomy. Subsequently, in 1932, the Italian Antonino Spinelli described the onset of this painful syndrome in athletes, especially footballers and fencers [7], thereby attracting the attention of the scientific community which has since produced numerous publications on the clinical diagnostic assessment and treatment of osteitis pubis.

Many conditions are associated with the development of osteitis pubis: not only pelvic surgery but also the abuse of parenteral drugs, pregnancy and childbirth, major trauma or repeated microtrauma to the pubic region and its muscle attachments as well as a variety of rheumatological conditions. Although the precise aetiology is still uncertain, recent studies seem to agree that the most likely cause is repeated microtrauma or shearing forces across the pubic symphysis [8]. In fact, the disorder is observed in athletes who engage in activities involving repetitive abnormal shearing forces across the symphysis, as in accelerations–decelerations or sudden changes of direction, which are, for example, common in football and all sports involving kicking [9]. The abdominal muscles act in synergy with the posterior paravertebral musculature to stabilise the symphysis, allowing the subject to maintain his balance while standing on one leg and therefore contributing to the strength and precision of the kick. Even the adductor muscles work to stabilise the symphysis, but they act as antagonists to the abdominal muscles and transmit traction forces towards the pubic symphysis during dodges or ball passes. Imbalances among these muscle masses alter the forces around the pubic symphysis, predisposing the athlete to subacute periostitis.

It has been suggested [10] that the antagonistic action between the longus adductor, which produces a shearing force in an anterior–inferior direction at the level of the symphysis, and the rectus abdominis, which pulls on the symphysis in a posterior–superior direction, may significantly affect the normal biomechanics of the pubic symphysis, creating strain and joint degeneration.

When the tissue's intrinsic regeneration capabilities are exceeded, tissue degeneration occurs. Moreover, abnormal shearing forces acting on the pubic symphysis can cause sacroiliac joint dysfunction thereby aggravating the pelvic biomechanical changes and pubic symphysis microtraumas and leading to inflammation and muscle contracture [11].

A correlation has been shown between pubic symphysis stress injury and degenerative changes of the sacroiliac joint, such as sacral stress fractures [12].

16.3 Anatomy

The anatomy of the pelvic girdle is quite complex. The pelvis presents as a ringlike structure in which any change to the anatomy or forces applied to it (e.g. due to different length of lower limbs or sacroiliac joint dysfunction) can lead to an increase in stress and pain.

The pubic symphysis is described in recent anatomy books as a secondary or fibrocartilaginous joint, definitions that have replaced the functional terms of synarthroses or amphiarthroses used in the classic textbooks. It is a stiff, or better a semimobile, joint capable of resisting high tensile, tangential and compressive forces and allowing only slight physiological articular movement (<2 mm) in the majority of adults. It is made up of two pubic bones presenting, on their medial aspect, an elliptical or oval facet, the long axis of which slants downwards and backwards. In the space between the two pubic bones there is a fibrocartilaginous disc, the interpubic disc, which, by closely adhering to the joint facets, guarantees most of the stability of the diarthrosis. It has a transverse anterior width of around 5–6 mm and an anterior–posterior width of 10–15 mm [13].

This structure has two different portions: a peripheral portion which is very dense and resistant and a central portion which is softer and less dense. Other differences are related to sex: the effects of hormones on elastin during pregnancy cause the interpubic disc, as well as the soft parts of the interpubic and sacroiliac joints, to become more elastic to aid distension of the pubic joint and descent of the fetus. As a result of this increased pelvic elasticity, pregnant women may suffer from groin pain, which may be exacerbated by standing, walking and climbing up and down stairs because alternating weight bearing from one leg to the other creates tangential stress on the symphysis. These symptoms do not necessarily resolve with childbirth and may prove debilitating for some women [14].

Moreover, interpubic fibrocartilage is related to the peripheral ligaments as it is continuous with them with no distinct boundary. The superior pubic ligament is made up of a fibrous slip that extends horizontally between the pubic tubercles. Inferiorly, it blends with the fibrous membrane and superiorly with the abdominal linea alba. The inferior pubic ligament, or arcuate ligament of the pubis, is composed of a very resistant fibrous lamina, in the shape of a crescent with posterior–inferior concavity, which extends between the inferior pubic rami. Its upper border is closely adherent to the fibrous membrane, and the lower border makes up the pubic arch (wider in women than in men). The pubic symphysis is reinforced by extensions of the aponeurosis of insertion of the rectus abdominis, pyramidalis, external oblique and internal oblique muscles and extensions of the tendons of the pectineus, gracilis, adductor longus, adductor brevis, external obturator and internal obturator muscles. All these interlacing fibres form a thick network in front of the anterior surface of the pubic symphysis [15].

The pubis is completely covered by periosteum with the exception of the articular facet with the pubic bone on the other side. The periosteum histologically consists of an outer or superficial layer, also called fibrous layer, made up of longitudinally aligned, parallel bundles of connective tissue, intermingled with elastic fibres. The inner or deep layer has an abundant, tightly knit elastic reticulum, interspersed with thin, more or less arciform connective tissue fibres arranged longitudinally and transversally, which penetrate the bone and are commonly known as Sharpey's fibres. In young individuals, the deep layer of the periosteum is rich in osteoblasts that form Ollier's subperiosteal osteogenic layer. During growth, this layer decreases progressively until it disappears in adults, in whom periosteum is essentially devoid of regenerative capacity. It is therefore easy to understand why any repeated trauma on periosteum can lead to exostosis originating from the subperiosteum and pain [7].

The pubic joint is richly innervated by the genitofemoral nerve (L1, L2) and pudendal nerves (S2, S3 and S4), and it receives both sympathetic and parasympathetic nerve supplies from the same lumbar and sacral levels. For this reason, symptoms may be referred to the dermatomes and myotomes supplied by L1, L2, S2, S3 and S4. The L1 and L2 dermatomes include the groin, the greater trochanter, the lumbar spine area and the anterior surface of the thigh. The L2 myotome includes the hip flexor and adductor muscles. The S2, S3 and S4 dermatomes include the perineum, the medial surface of the thigh, the sacral and gluteal region and the groin area. The myotomes include external rotator muscles of the hip and the gluteal muscles [13].

16.4 Signs and Symptoms

The signs and symptoms associated with osteitis pubis are varied and nonspecific so that in many cases the condition is not readily recognised and treated: some studies report an average diagnostic delay of 3 months, with a range between 1 week and 1 year [13].

Typically, osteitis pubis occurs with pain in the inguinal region and lower abdomen that may radiate to the groin, genitals, perineum, thigh or hip. Onset is usually subtle with patients presenting days or even weeks after the appearance of pain; however, in some cases, onset is sudden. The most common presentations include:

- Pubic pain that radiates outwards, especially felt when the subject is standing on one leg, often described as a burning sensation in the groin or in other cases as a sharp or stabbing pain
- Pain at the adductor musculature insertion on the symphysis pubis, frequently unilateral
- Pain exacerbated by dynamic stresses such as running, kicking, side-to-side movements, accelerations and decelerations, leaps, pivoting on one leg and lying on the affected side
- Pain occurring with plain activities like walking, climbing stairs, coughing and sneezing and that disappears with rest
- A sensation of clicking at the pubic-coxofemoral level when sitting up, turning over in bed and walking on uneven surfaces
- Weakness and difficulty walking
- Sometimes fever possibly associated with chills (in such cases, osteomyelitis must be ruled out especially if the patient has undergone pelvic surgery) [8]

The objective signs of osteitis pubis may vary widely and include:

- Tenderness on palpation of the area overlying the superior pubic ramus
- Pain over one or both sacroiliac joints, frequently associated with piriformis spasm and consequent sciatic-like pain
- In the event of different-length legs, coxofemoral pain in the longer limb [8]

Complications of osteitis pubis are uncommon and include injury to the adductor muscle tendons and, if the condition is misdiagnosed, progression to bone erosion. Cases of femoral artery involvement have also been reported, but the incidence is extremely low.

16.5 Diagnosis

The diagnosis of osteitis pubis is essentially clinical.

A thorough history and a careful physical examination are fundamental. The range of motion of one or both hips may be reduced, in particular in abduction, due to spasm of the antagonist musculature.

One of the various possible tests is the single-leg hop test, which may reproduce the patient's symptoms, and having the patient place his or her fists between the knees and press against them while maintaining a prolonged contraction of the adductor muscles. The most specific, albeit simple, test for the diagnosis of osteitis pubis is the so-called direct-pressure spring test which involves palpating over the symphysis pubis to elicit tenderness to touch. Sliding the fingertips a few centimetres laterally and applying direct pressure on the pubic ramus will elicit pain in a patient with osteitis pubis [8]. Examination of the patient must also include an assessment of gait and of the strength of the adductor and flexor muscles.

At this stage of the assessment, it is important to rule out other pathological conditions that may give rise to a similar clinical picture to osteitis pubis. These include:

- Gynaecological conditions (pelvic inflammatory disease or extrauterine pregnancy)
- Urological conditions (prostatitis, prostatic cancer, urinary tract infection)
- Inguinal hernias
- Bursitis
- Tendinopathy
- Acetabular labral tears
- Muscle strains or contusions
- Pelvic or coxofemoral fractures
- Osteomyelitis
- Maigne syndrome (rare)
- Reiter syndrome
- Psoriatic arthritis
- Ankylosing spondylitis [8]

Laboratory investigations are not usually necessary but may nonetheless help to rule out other diagnoses. These include blood tests (ESR and WBC), urinalysis and possible blood culture in febrile patients.

Imaging studies appear more useful. Plain radiography may be totally negative in the early

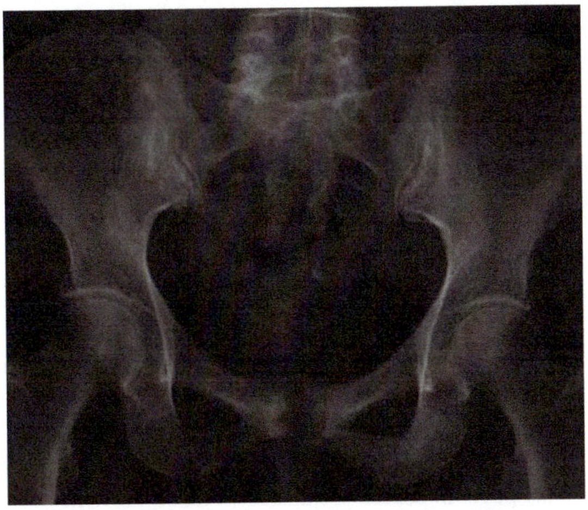

Fig. 16.1 Plain radiograph of the pelvis in a patient with osteitis pubis shows the classical findings of reactive sclerosis, widening of the symphysis pubis, bone thinning, osteolysis and joint irregularities

stages of the disorder, but after a few weeks (about 4 according to recent studies), there may be widening of the symphysis pubis, followed by possible sclerosis, osteolysis and bone erosions (Fig. 16.1). Sometimes there may also be evidence of pubic instability, defined as a cephalad translation greater than 2 mm of the superior pubic ramus, with the patient standing on one leg. The sacroiliac joint may also be evaluated.

Bone scintigraphy with technetium-99m or single-photon emission computed tomography (SPECT) may also be negative in early osteitis pubis but become positive over time and show increased tracer uptake directly over the pubic symphysis. However, there is no correlation between the degree of positivity on imaging and the severity of symptoms.

A computed tomography (CT) scan may also be requested to evaluate the pubic symphysis and the posterior pelvic ring.

A growing role is played by magnetic resonance imaging (MRI) (Fig. 16.2), especially when performed with fat-suppressed sequences. Typical findings of osteitis pubis of less than 6 months' duration include subchondral bone oedema, periarticular effusion and oedema. Some of these findings may also be detected in asymptomatic patients. Osteitis of more than 6 months'

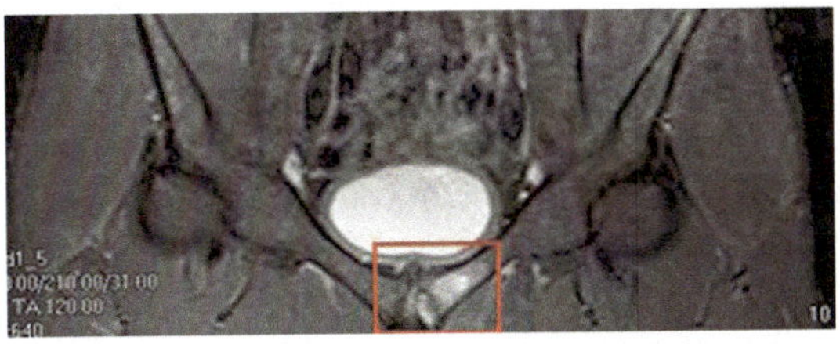

Fig. 16.2 Coronal MRI of a patient with osteitis pubis. The *boxed area* shows bone oedema, a finding which, combined with the clinical findings, can help define the diagnosis

duration will typically show subchondral sclerosis, bone resorption and irregularity of the bone margins and osteophytes [16].

16.6 Treatment

The treatment of osteitis pubis is essentially conservative. The main treatment goals are to relieve pain and identify and correct the underlying biomechanical imbalances. The mainstay of treatment is rest, which can be combined with physical treatments, especially useful in early stages, and the scrupulous administration of nonsteroid and steroid anti-inflammatory drugs. Initially the athlete will be instructed to refrain from sport activities, as the condition is self-limiting and generally resolves in 6–12 months. Mean duration of prescribed rest ranges from 2 weeks to 3 months, depending on the case. In the acute phase, physical therapies such as ultrasound or TENS may be useful, and alternating ice and heat may help ease the pain. In the early phase, it is useful to prescribe ambulation with an assistive device (e.g. a cane or crutches) and possible orthoses (e.g. a lumbar corset) in the presence of concurrent lumbosacral overload or sacroiliac joint disorder.

In the acute phase, the patient will benefit from physical therapy in water to preserve mobility of the pelvic girdle and coxofemoral joints in a controlled environment. During the acute phase, nonsteroid anti-inflammatory drugs can be prescribed to reduce the pain and inflammation. Less frequently, if the symptoms persist, steroid injections into the pubic symphysis may be taken into consideration [13].

Once the pain has subsided, the patient can start strengthening therapy, with specific exercises for the hip flexors and hip adductors, abdominal muscles and lumbar stabilizers. Stretching exercises for the hamstrings and femoral quadriceps are also commonly included.

A treatment strategy focusing on a programme of active exercises proves to be superior compared to an approach based on physical therapies such as laser therapy, deep transverse massage and TENS. The use of exercises to improve neuromuscular coordination, strengthen the pelvic girdle musculature and strengthen any muscle groups found to be weak on clinical kinesiological tests, such as the gluteus medius, iliopsoas and abdominal muscles, has shown to be more effective, as demonstrated by a prospective randomised trial [17].

In some cases, swimming may be advised, with the exception of the breaststroke to avoid excessive use of the adductors. Manipulations may also alleviate pain, relax muscle spasms and increase flexibility of the pubic symphysis.

Return to running should be gradual and well managed, decreasing the intensity and duration of activity if the symptoms reappear.

Surgical treatment is rare and reserved for those few cases in which conservative treatment fails. Various surgical techniques may be performed including curettage, arthrodesis and varying extents of bone resection. Surgery has been associated with an initial improvement of symptoms, but it may lead to posterior pelvic instability in the long term [6, 7].

Return to play is recommended only when the patient is pain-free.

The prognosis of osteitis pubis is excellent, although in some cases the condition has a tendency to recur. Fundamental measures to prevent recurrence are an adequate training programme with daily sessions to preserve the normal flexibility of the pubic symphysis and sacroiliac joint, the use of adequate equipment and shoes and the early recognition of symptoms to enable timely and appropriate intervention.

References

1. Volpi P, Melegati G. La pubalgia del calciatore. Aspetti eziopatogenetici e classificativi. Ital J Sports Traumatol. 1985;8:271–4.
2. Fricker P, Taunton JE, Ammann W. Osteitis pubis. Infection, inflammation or injury? Sports Med. 1991;12(4):266–79.
3. Orchard J, Seward H. Epidemiology of injuries in the Australian Football League, season 1997–2000. Br J Sports Med. 2002;2(1):39–44.
4. Lovell G. The diagnosis of chronic groin pain in athletes: a review of 189 cases. Aust J Sci Med Sport. 1995;27(3):76–9.
5. Cunningham PM, Brennan D, O'Connell M, et al. Patterns of bone and soft-tissue injury at the symphysis pubis in soccer players: observation at MRI. Am J Roentgenol. 2007;188(3):291–6.
6. Beer E. Periostitis of the symphysis and descending rami of the pubes following suprapubic operations. Int J Med Surg. 1924;37(5):224–5.
7. Spinelli A. Una nuova malattia sportiva: la pubalgia degli schermidori. Orthop Trauma App Mot. 1932;4:111.
8. Goitz HT. Osteitis pubis. Medscape reference. eMedicine. http://emedicine.medscape.com/article/87420-overview. Updated Dec 29, 2015. Accessed [+data ultimo accesso].
9. Andrews SK, Carek PJ. Osteitis pubis: a diagnosis for the family physician. J Am Board Fam Pract. 1998;11:291–5.
10. Omar IM, Zoga AC, Kavanagh EC, et al. Athletic pubalgia and "sports hernia": optional MR imaging technique and findings. Radiographics. 2008;28(5):1315–438.
11. Rodriguez C, Miguel A, Lima H, Heinrichs KI. Osteitis pubis syndrome in the professional soccer athlete: a case report. J Athl Train. 2001;36(4):437–40.
12. Major NM, Helms CA. Pelvic stress injuries: the relationship between osteitis pubis (symphysis pubis stress injury) and sacroiliac abnormalities in athletes. Skeletal Radiol. 1997;26(12):711–7.
13. Holt MA, Keene JS, Graf BK, Helwig DC. Treatment of osteitis pubis in athletes. Results of corticosteroid injections. Am J Sports Med. 1995;23(5):601–6.
14. Owens K, Pearson A, Mason G. Symphysis pubis dysfunction—a cause of significant obstetric morbidity. Eur J Obstet Gynecol Reprod Biol. 2002;105:143–6.
15. Kapandji IA. The physiology of the joints, Trunk and vertebral column, vol. Vol. 3. 2nd ed. Edinburgh: Churchll Livingstone; 1974. p. 53–67.
16. Kunduracioglu B, Yilmaz C, Yorubulut M, Kudas S. Magnetic resonance findings of osteitis pubis. J Magn Reson Imaging. 2007;25:535–9.
17. Holmic P, et al. Effectiveness of active physical training as treatment for long-standing adductor-related groin pain in athletes: randomized trial. Lancet. 1999;353:439–43.

Part 6

Neurologic Pathology

Nerve Entrapment Syndromes

17

Giancarlo Barolat

17.1 Anatomy

The main nerves innervating the inguinal/groin region include the ilioinguinal, iliohypogastric, genitofemoral, and obturator nerves. Globally they can be referred to as the "inguinal nerves."

17.1.1 Ilioinguinal and Iliohypogastric Nerve

The ilioinguinal and iliohypogastric nerves sometime arise as a common trunk and, in their course, usually separate between the transversus and internal oblique muscles.

Iliohypogastric nerve. The iliohypogastric nerve arises primarily from the ventral primary rami of L1 and occasionally with a twig from T12. This nerve has a pathway similar to that of the intercostal nerves in the thoracic region. The iliohypogastric nerve traverses the psoas major muscle, piercing the lateral border of the muscle anterior to the quadratus lumborum muscle and posterior to the kidney to traverse the lateral abdominal wall. Superomedial to the anterior superior iliac spine, the iliohypogastric and ilioinguinal nerves pierce the transversus abdominis to lie between it and the internal oblique muscles. After traveling a short distance inferomedially, their ventral rami pierce the internal oblique to lie between the internal and external oblique muscles before giving off branches, which pierce the external oblique to provide cutaneous sensation. The iliohypogastric nerve supplies the skin over the inguinal region as well as a small region just superior to the pubis.

Ilioinguinal nerve. The ilioinguinal nerves emanate from the first lumbar spinal root. In one study of 200 human bodies, the ilioinguinal nerve arose from the lumbar plexus in 72.5% and by a common trunk with the iliohypogastric nerve in 25%; it was absent in 2.5% (Fig. 17.1). The ilioinguinal nerve was formed from one root in 92.5% and from two roots in about 5% of cases. In 86%, the ilioinguinal carried fibers from one spinal nerve (primarily from L1) and, in 11%, from two spinal nerves (T12, L1; L1, L2; or L2, L3) [1].

Within the inguinal canal, the nerve usually lies ventral to the spermatic cord (60% of cases), but it may lie beneath (dorsal) the cord and/or within it. The ilioinguinal nerve runs anteroinferiorly to the superficial inguinal ring, where it emerges to supply the skin on the superomedial aspect of the thigh. The ilioinguinal nerve, usually smaller than the iliohypogastric nerve, arises with it from the first lumbar nerve. It emerges from the lateral border of the psoas major just below the iliohypogastric and, passing obliquely across the quadratus lumborum and iliacus, perforates the transversus abdominis, near the

G. Barolat, M.D.
Barolat Neuroscience, Presbyterian/St. Lukes Medical Center, Denver, CO, USA
e-mail: gbarolat@gmail.com

R. Zini et al. (eds.), *Groin Pain Syndrome*, DOI 10.1007/978-3-319-41624-3_17

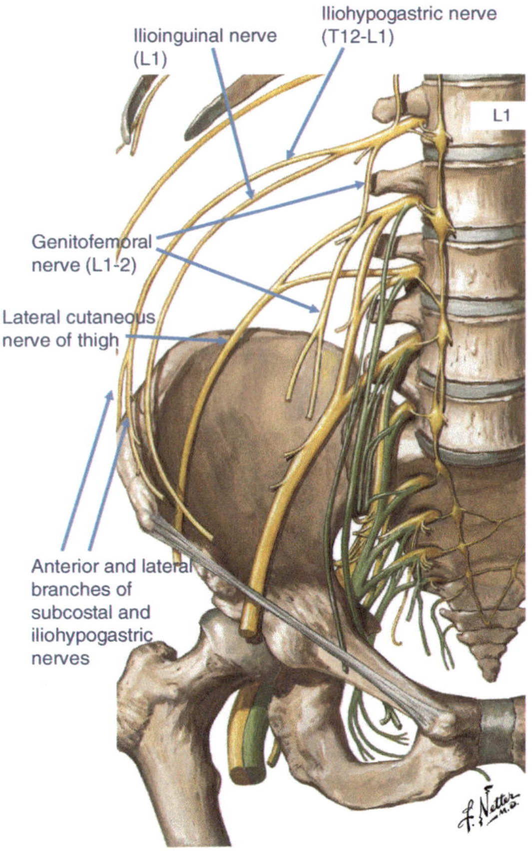

Fig. 17.1 Origin of "inguinal nerves" from the lumbar plexus

anterior part of the iliac crest, and communicates with the iliohypogastric nerve between the transversus and the obliquus internus. The nerve then pierces the obliquus internus and, accompanying the spermatic cord through the subcutaneous inguinal ring, is distributed to the skin of the upper and medial part of the thigh, to the skin over the root of the penis and upper part of the scrotum in the male; in the female it provides sensory innervation to the skin covering the mons pubis and labium majus in the female (Fig. 17.2). The size of this nerve is inversely proportional to that of the iliohypogastric nerve. Occasionally, it is very small and ends by joining the iliohypogastric nerve; in these instances, a branch from the iliohypogastric might take the place of the ilioinguinal, or the latter nerve may be altogether absent. The ilioinguinal nerve may partially or completely replace the genital branch of the genitofemoral nerve or the lateral femoral cutaneous nerve.

Of note, the ventral rami of the lower intercostal nerves (T11 and T12) also pierce the transversus abdominis muscle to lie between it and the internal oblique. These latter nerves also supply sensation to the inferior abdominal wall.

17.1.2 Genitofemoral Nerve

The genitofemoral nerve refers to a human nerve that is found in the abdomen. Its branches, the genital branch and femoral branch, supply sensation to the upper anterior thigh, as well as the skin of the anterior scrotum in males and mons pubis in females. The femoral branch is different from the femoral nerve, which also arises from the lumbar plexus. The genitofemoral nerve originates from the upper L1-2 segments of the lumbar plexus. It passes downward and emerges from the anterior surface of the psoas major muscle. The nerve continues downward and divides into two branches, the genital branch and the femoral branch. The genital branch passes through the deep inguinal ring and enters the inguinal canal. In men, the genital branch continues down and supplies the scrotal skin. In women, the genital branch accompanies the round ligament of uterus, terminating in the skin of the mons pubis and labia majora.

The femoral branch passes underneath the inguinal ligament, traveling adjacent to the external iliac artery and supplying the skin of the upper, anterior thigh.

17.1.3 Obturator Nerve

The obturator nerve originates from the ventral divisions of the ventral rami of the L2 through L4 spinal nerves within the psoas major muscle. The obturator nerve descends through the psoas muscle to emerge from its medial border at the pelvic brim. The nerve then curves downward and forward, around the wall of the pelvic cavity,

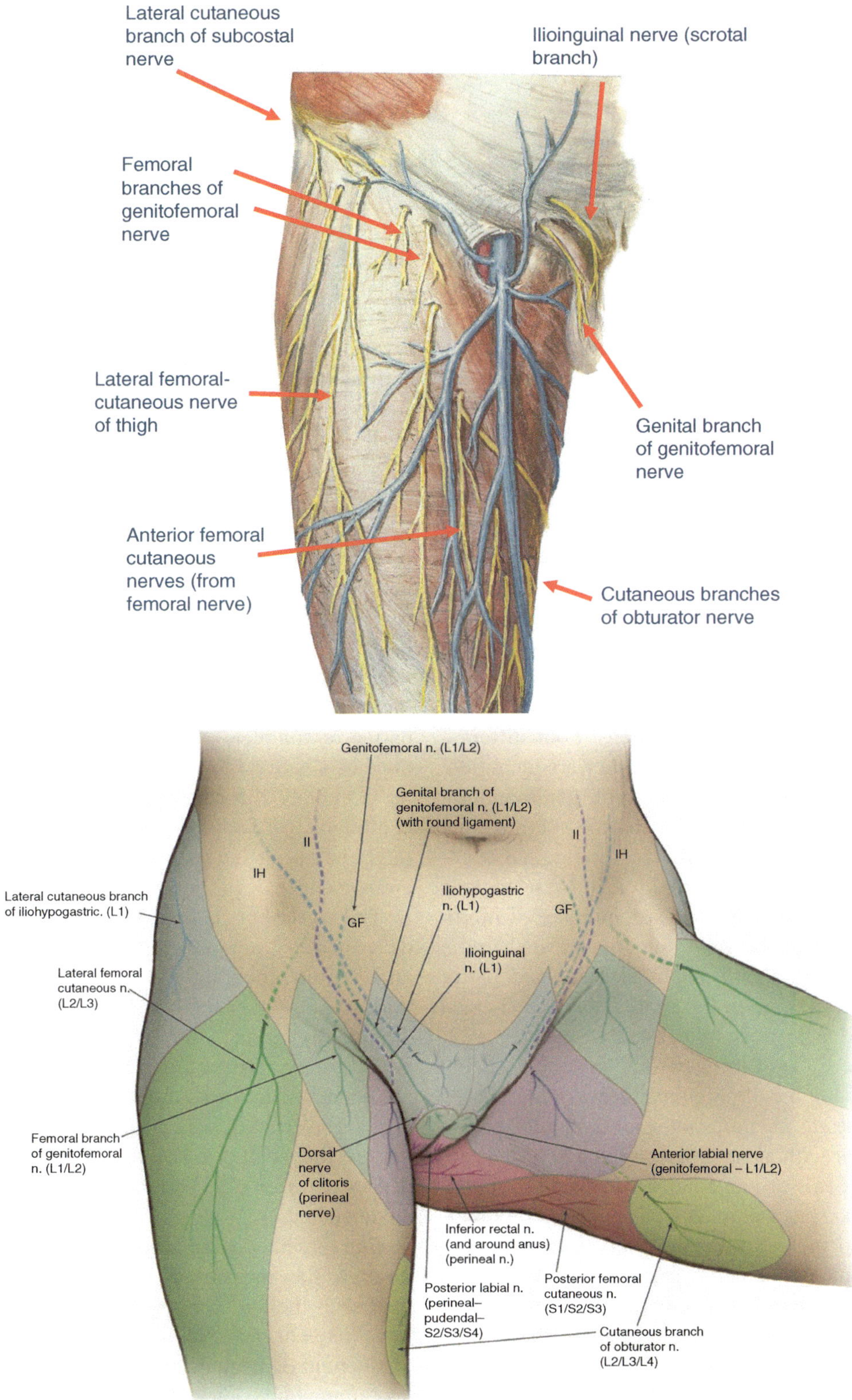

Fig. 17.2 Cutaneous distribution of the branches of the "inguinal nerves"

and travels through the obturator foramen in which it divides into the anterior and posterior branches.

17.1.4 Lateral Femoral Cutaneous Nerve

The lateral femoral cutaneous nerve is a branch of the lumbar plexus, exiting the spinal cord between the L2 and L3 vertebrae. It emerges at the lateral edge of the psoas muscle group, below the ilioinguinal nerve, and then passes beneath the iliac fascia and the inguinal ligament. It divides into two branches—anterior and posterior—8–10 cm below the spine, where it also emerges from the fascia lata. The anterior branch innervates the skin of the anterior and lateral regions of the thigh to the knee; the posterior branch supplies the lateral portion of the thigh, from the greater trochanter in the hip to mid-thigh, just above the knee.

Rab et al. [2] conducted an anatomical study of the groin nerves in 64 halves of 32 human embalmed anatomic specimens. Four different patterns of cutaneous branching were identified: type A, with a dominance of genitofemoral nerve in the scrotal/labial and the ventromedial thigh region. In type A, the ilioinguinal nerve gives no sensory contribution to these regions (43.7%). In type B, with a dominance of ilioinguinal nerve, the genitofemoral nerve shared a branch with the ilioinguinal and gave motor fibers to cremaster muscle in the inguinal canal but had no sensory branch to the groin (28.1%). In type C, with the genitofemoral nerve being dominant, the ilioinguinal nerve had sensory branches to the mons pubis and inguinal crease together with an anteroproximal part of the root of the penis or labia majora. In type D, cutaneous branches stemmed from both the ilioinguinal and the genitofemoral nerves. In addition, the ilioinguinal nerve provided innervation to the mons pubis and inguinal crease together with a very anteroproximal part of the root of the penis or labia majora (7.8%). These patterns of innervation were bilaterally symmetric in 40.6% of the cadavers.

Al-dabbagh [3] conducted a study in a consecutive series of 110 primary inguinal hernias repaired by the mesh technique. Particular attention was paid to early identification and recording of the course of both the ilioinguinal and iliohypogastric nerves and preserving them throughout the operative procedure. The course of both nerves was found to be consistent with that described in anatomical texts in only 41.8% of the surgical explorations. The course of one or both nerves was found to be a variant in the other 64 of 110 (58.2%) explorations and often rendered them susceptible to injury.

17.2 Clinical

It is important to recognize the difference between conditions where the nerve is subjected to persistent compression (which can eventually lead to permanent damage) versus condition where the nerve has been intrinsically damaged (and the damage can be temporary or permanent, partial or complete). Sometimes this differential diagnosis is feasible; in other instances, it might be extremely difficult to differentiate the two conditions, particularly in situations where both might be coexisting (nerve damage and ongoing entrapment).

The different conditions would be best described as "inguinal entrapment neuropathy" and "inguinal neuralgia." The symptomatology may start in the exact distribution of one of the "inguinal region" nerves but then spread beyond their territory, including, at times, the abdomen or the whole thigh. Since there is significant overlap in the distribution of the groin nerves, the exact contribution of the individual nerves to the pain might not be clear in advanced cases.

Entrapment occurs where the nerve passes through the muscles of the abdominal wall medial to the anterior superior iliac spine. The most frequent cause of ilioinguinal neuropathy is inadvertent damage during herniorrhaphies and, less commonly, appendectomies [4, 5]. Other operations that may be associated with this neuropathy include tubal ligation, hysterectomy, and cesarean section. Other pathological conditions in the inguinal canal that can involve either the ilioinguinal or genitofemoral nerves are lipomas, leiomyomas, and endometriosis. Injuries to the iliohypogastric nerve mainly occur if the incision extends beyond the lateral margin of the inferior rectus abdominis fibers. Nerve(s) damage can result from direct surgical trauma, such as passing a suture around the nerve and incorporating it into the fascial repair, or postoperative entrapment in scar tissue or neuroma formation. Sports injuries, such as trauma or muscle tears of the lower abdominal muscles, may also result in injury to the nerves [6]. Injury to the iliohypogastric nerve may also occur during pregnancy, owing to the rapidly expanding abdomen in the third trimester. This is called the idiopathic iliohypogastric syndrome and is rare.

Neuralgia symptoms include burning or lancinating pain immediately following the abdominal operation. The pain extends from the surgical incision laterally into the inguinal region and suprapubic region.

Pain, numbness, and paresthesiae secondary to damage to the ilioinguinal nerve involve the upper media thigh, the base of the penis and scrotum in men, and the labia majora and mons pubis in women. Discomfort may occur immediately or up to several years after the procedure and may last for months to years. The pain could be permanent. This discomfort is possibly because of the formation of scar tissue in the region. Occasionally, the pain may extend into the genitalia because of significant overlap with other cutaneous nerves.

Rarely, ilioinguinal neuralgia will occur spontaneously. Many patients with an ilioinguinal neuropathy notice minimal numbness in the distribution of the nerve that resolves during days or weeks, but others complain of more severe pain. This syndrome has been called post-herniorrhaphy neuralgia, inguinal neuralgia, and genitofemoral and ilioinguinal neuralgia. It might be preferable to refer to it as inguinal neuralgia because a specific nerve is not implicated. Ilioinguinal neuralgia is one of the most common causes of lower abdominal and pelvic pain encountered in clinical practice. These patients have severe burning and stabbing pain in the lower abdomen, inguinal area, and upper thigh often aggravated by changing position and walking. On exam, if the neuropathy has been of sufficient intensity for a long enough period, there will be an alteration of sensation in this cutaneous zone. This could result in hypoesthesia as well as hyperesthesia. In extreme cases, there is allodynia, which consists of extreme pain even to light touch. Tinel's sign may be elicited by tapping over the ilioinguinal nerve at the point at which it pierces the transverse abdominal muscle. The pain of ilioinguinal neuralgia is made worse by extension of the lumbar spine, which puts traction on the nerve. Patients suffering from ilioinguinal neuralgia will often assume a bent-forward novice skier's position. If the condition remains untreated, progressive motor deficit consisting of bulging of the anterior abdominal wall muscles may occur. This bulging may be confused with inguinal hernia.

The symptoms in genitofemoral neuropathy/algia usually consist of an unpleasant painful feeling in the lower abdomen and groin, with pain radiating to the inner side of the upper leg, scrotum, or greater labia.

Injuries to the obturator nerve are uncommon. Pelvic fractures are an obvious cause. The obturator nerve can be damaged due to hip surgery or by pelvic malignant neoplasms or hernias, foci of endometriosis. Pain typically is localized to the adductor origin at the pubic bone [7].

17.3 Management

Initial treatment of inguinal neuropathy/algia should consist of simple analgesics, nonsteroidal anti-inflammatory drugs, membrane-stabilizing medications (carbamazepine, gabapentin, pregabalin), and/or tricyclic antidepressant drugs. Opioid medications are seldom indicated at this stage. Avoidance of repetitive activities thought to exacerbate the symptoms of ilioinguinal neuralgia (e.g., squatting or sitting for prolonged periods) will also help ameliorate the patient's symptoms.

If the patient fails to respond to these conservative measures, a next reasonable step is nerve blockade with local anesthetic and steroid.

Because of the overlapping innervation of the ilioinguinal and iliohypogastric nerve, it is not unusual to block branches of each nerve when performing ilioinguinal nerve block. The clinician should be aware that due to the anatomy of the ilioinguinal nerve, damage to or entrapment of the nerve anywhere along its course can produce a similar clinical syndrome.

For patients who do not rapidly respond to nerve(s) block, or whose pain has spread beyond the mere distribution of an individual nerve, consideration should be given to epidural steroid injection of the T12-L1 segments intraspinally.

Repeated nerve blocks can be given, and sometimes they can result in cure from the condition.

If nerve blocks are effective but their efficacy is short-lived, pulsed radiofrequency lesioning (PRF) of the involved nerves might obtain long-term results [8]. Werner et al. performed a systematic review of the available data on PRF ablation in the treatment of chronic post-herniorrhaphy inguinal pain [9]. They report that there is a limited level of evidence to support the use of PRF ablation in the management of this condition, stating that the evidence is of low quality and the strength of recommendation is weak to moderate.

If the above treatments have not resulted in any improvement, a reassessment of the situation is in order.

Namely, one has to decide whether this is a nociceptive, curable pain, or whether this is a permanent neuropathic pain (Fig. 17.3). Persistent nociceptive pain is a reflection of a nerve still affected by ongoing damage by pressure or constriction (scar tissue, mesh, suture, etc.) or whether this constitutes a "neuropathic" pain, a reflection of an intrinsically permanently and irreversibly damaged nerve. To make matters more complicated, when "neuropathic" pain persists for a long time, it might become "centralized," which means that alternative pain-generating/self-maintaining circuits originate in the central nervous system. To complicate the situation even further, sometimes both nociceptive and neuropathic pain can be present concurrently, and sometimes they can be very difficult to be teased apart clinically. In the author's experience, three clinical signs/symptoms are almost always indicative of neuropathic pain. The first is the fact that the pain is usually present 24 h/day. In most instances of nociceptive pain, the pain is greatly lessened by rest. Secondly, neuropathic pain is usually described as a burning sensation. Thirdly, the presence of allodynia (pain even to light touch) is almost always associated with neurogenic pain. Therefore, if a patient is complaining of constant burning pain and displays allodynia, it is safe to assume that at least a component of the pain is neuropathic and most likely "centralized."

The differential diagnosis is important, since the goals of treatment are drastically different. In the presence of nociceptive pain, one should try to "cure" the condition. Neuropathic pain cannot be cured, but only managed.

If one is aiming for the treatment of nociceptive pain, surgical intervention might be indicated. Different surgical approaches can be considered.

A decompression of a possibly entrapped nerve can be performed. This might entail lysis of adhesions or take down of scar tissue and/or removal of implanted hardware such a mesh or retained sutures. Aasvang and Kehlet [10] performed a literature review for data on surgical treatment of chronic pain after hernia repair. In some patients, pain may have been of inflammatory or nerve entrapment origin from the mesh, suggesting the need for mesh removal or nerve decompression. However, there were no data on how to diagnose these conditions and few on the success of mesh removal; only four patients with mesh removal alone were reported. Lee and Dellon [11] reported on their surgical management of 54 patients with groin pain history and physical examination were sufficient to relate the pain to one or more of the lateral femoral cutaneous, ilioinguinal, iliohypogastric, or genitofemoral nerves. In their series,

Nociceptive pain	• Nociceptors are the nerve endings which sense and respond to parts of the body which suffer from damage. They signal tissue irritation, impending injury, or actual injury. • Nociceptive pain is due to activation of nociceptors • It is the "usual pain" that follows tissue injury • Usually time limited • Usually improves as tissues heal • Often with an aching or throbbing quality • Usually improves with rest
Neuropathic pain	• Due to damage to the nervous system • Sometimes becomes permanent • Issue is a "faulty electrical circuitry" within the nervous system • Can seldom be cured • Often present 24 hours/day • Often worse at night • Burning is pathognomonic of neuropathic pain • Allodynia (pain to light touch) is pathognomonic of neuropathic pain • If persistent, usually pain signals no longer represent an alarm about ongoing or impending injury, instead the alarm system itself is malfunctioning

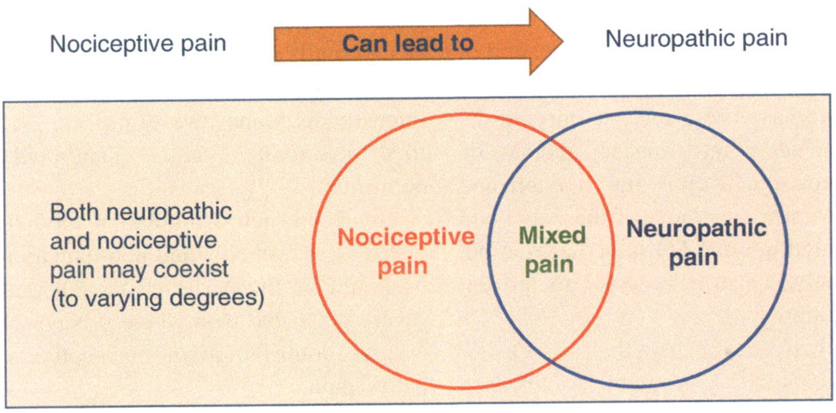

Management

Nociceptive pain	• Try to correct the underlying cause • Analgesic and anti-inflammatory medications • Physical therapy • Possible nerve decompression if entrapment neuropathy is suspected
Neuropathic pain	• Membrane-stabilizing medications • Nerve blocks • Pulsed radio-frequency or cryoablation • Neurectomy • Neurostimulation If patient has a clearly centralized pain syndrome (pain 24 hours/day + burning pain + allodynia + pain extending beyond distribution of affected nerves) neurostimulation should be preferred over neurectomy

Fig. 17.3 Neuropathic vs. nociceptive pain. Diagnostic and therapeutic considerations

neurolysis was performed in all instances of lateral femoral cutaneous nerve entrapment. Ninety percent of those patients experienced good to excellent results. Entrapments of the other three nerves were addressed by resection. Pain related to the genitofemoral nerve had the worst outcome.

Often a resection or sectioning of the nerve(s) is performed [12, 13]. Some authors have reported excellent results after sectioning of the three nerves involved (ilioinguinal, iliohypogastric, and genitofemoral nerves), called the "triple neurectomy" procedure [14–16]. This can be performed either laparoscopically or via an open technique. Madura et al. [17] reported on a consecutive series of 100 patients who underwent inguinal neurectomy for inguinal nerve entrapment. Seventy-two percent of the patients obtained good long-term relief. The authors emphasized the need to address the distal end of the resected nerves, in order to prevent formation of painful neuromas. These include burying the cut end of the nerve into muscle, end-to-side nerve anastomosis, and epineural ligation and flap. Experimentally, treatment of the nerve end with electro-fulguration, YAG laser destruction, and tissue bioglues also is reported to prevent neuroma formation.

One should beware of the fact that, if the original pain is exquisitely neuropathic (defined as excruciating constant burning pain and allodynia), these surgical ablative approaches can result in a severe (at times permanent) worsening of the pain.

In the author's experience, if the pain is clearly neuropathic, the condition is a reflection of an "electrical" problem within the nervous system. In these instances, electrical stimulation through an implantable neurostimulation device can often be very effective in reducing the pain signals and can actually represent the most effective treatment option [18–23]. Certainly, it is the most reversible surgical modality and the one that would cause the least irreversible damage to the nervous system. Electrical signals are delivered through implanted electrodes. The electrodes can be implanted on the peripheral nerves, on the small subcutaneous nerve fibers of the affected painful area, or on the nerve roots or dorsal root ganglia (DRG) serving that area (usually T12-L1, Figs. 17.4, 17.5, and 17.6). The stimulation can be delivered as a test trial on a temporary basis. The electrodes are usually placed for about 7–15 days and then removed. If the patient experienced good pain relief, a complete neurostimulation system can then be implanted. This usually consists of a pulse generator and one or more electrodes. The pulse generator, very much like a pacemaker, contains a lithium battery (which can be externally rechargeable) and the electronic circuitry. The parameters of stimulation (polarity, intensity, frequency, waveform, etc.) can be modified externally via a hand-held wireless controller.

Since the neurostimulation modality can be tested in a reversible and non-damaging manner, it should be the preferred modality in cases of severe persistent pain where it is not clear of the relative contribution of nociceptive vs. neuropathic pain.

Neurostimulation can be effective for a very long time, and the author has a number of patients that were implanted 30 years ago and still maintain excellent pain relief.

Patients with chronic severe pain often develop severe depression, anxiety, and other mood disturbances. Psychological and psychosocial intervention is often of crucial importance in the support and management of these unfortunate individuals.

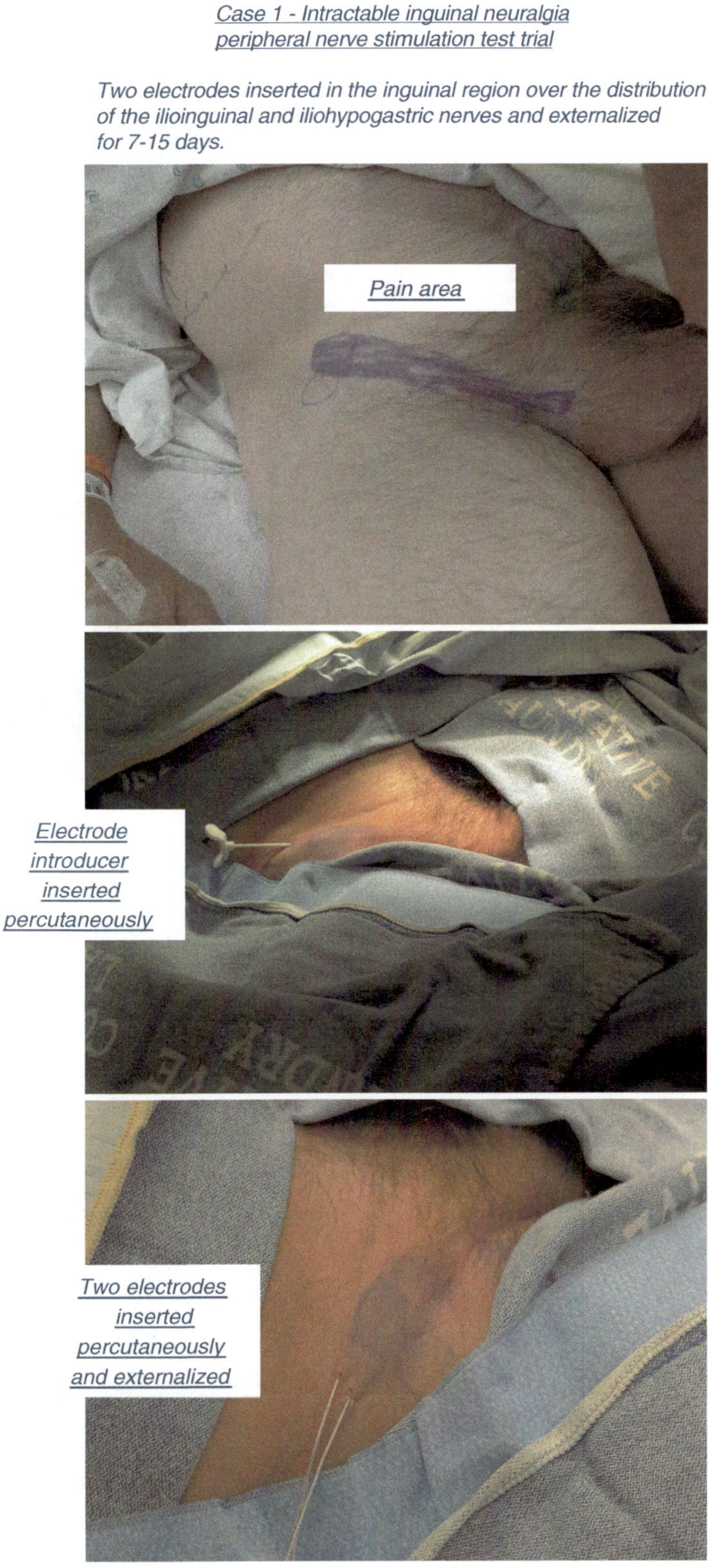

Fig. 17.4 Intraoperative images of insertion of two percutaneous leads over the ilioinguinal and iliohypogastric nerve distribution for a temporary test trial

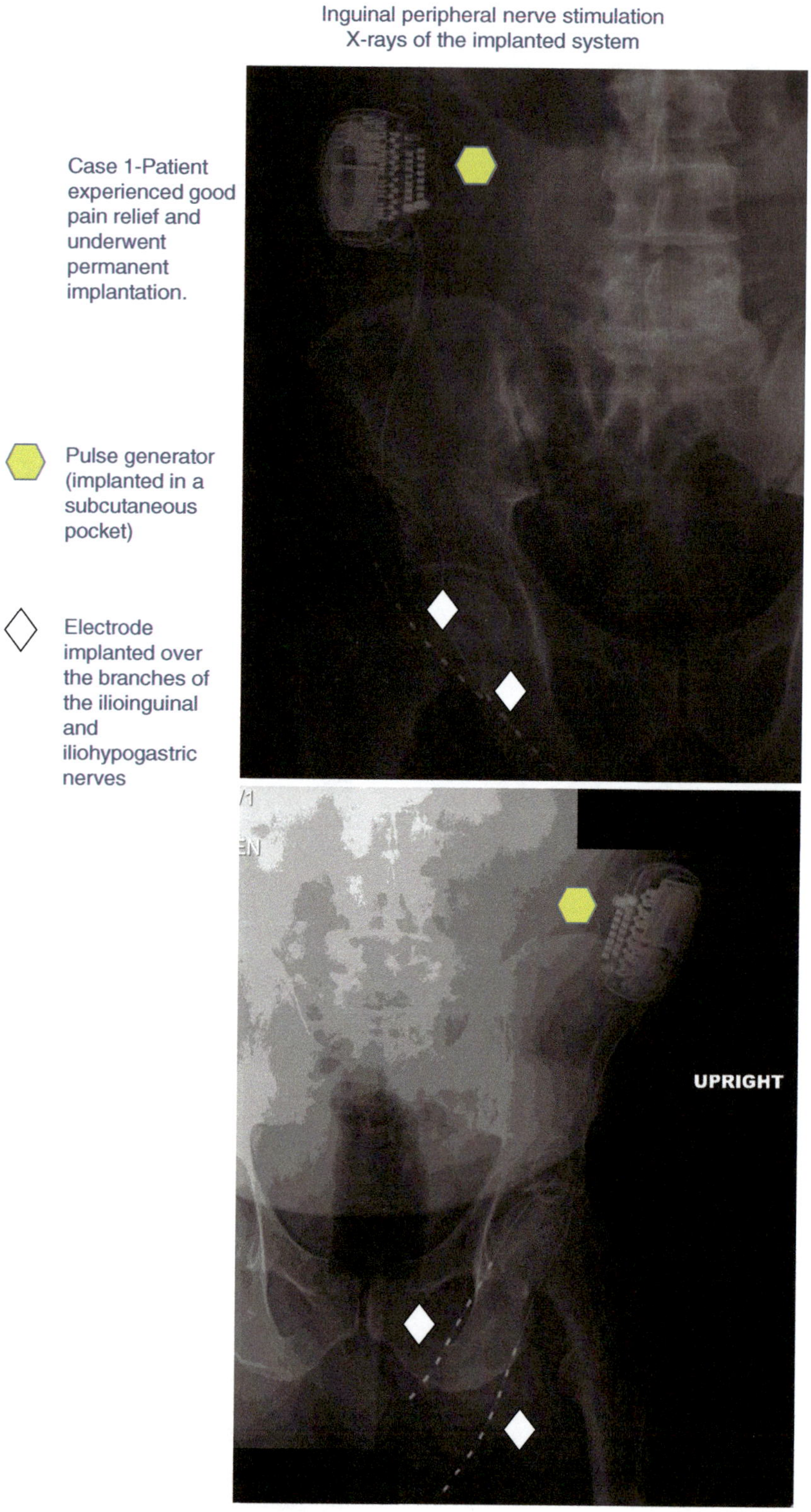

Fig. 17.5 X-rays of implanted peripheral nerve stimulation systems for inguinal pain

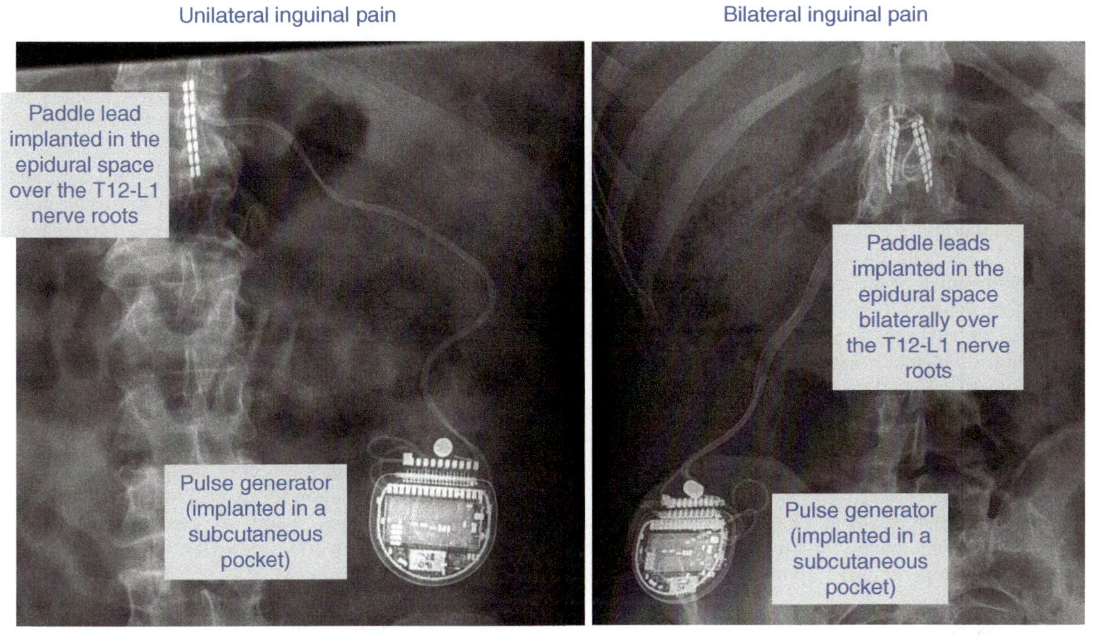

Fig. 17.6 X-rays of implanted T12-L1 dorsal nerve root stimulation systems for inguinal pain

Conclusions

Persistent neuropathies of the nervous structures innervating the groin can present a formidable challenge. The clinical spectrum can vary from some minimal transient numbness and discomfort to life-altering permanent excruciating pain conditions. Often the exact mechanisms and contribution of different nervous structures are ill-defined. In mild cases, a stepwise approach as described above is usually successful in providing meaningful control over the condition, which is often self-limited. In severe and chronic cases, the physician must be aware of the presence of neuropathic pain, which can make all the efforts to "cure" the condition doomed to fail. In those instances, membrane-stabilizing drugs and/or neurostimulation can provide meaningful, long-lasting relief. Neurectomy, while effective in the management of chronic nerve entrapment pain, is usually not effective in cases with "centralized" neuropathic pain.

References

1. Klaassen Z, Marshall E, et al. Anatomy of the ilioinguinal and iliohypogastric nerves with observations of their spinal nerve contributions. Clin Anat. 2011;24:454–61.
2. Rab M, Ebmer J, et al. Anatomic variability of the ilioinguinal and genitofemoral nerve: implications for the treatment of groin pain. Plast Reconstr Surg. 2001;108(6):1618–23.
3. Al-dabbagh A. Anatomical variations of the inguinal nerves and risks of injury in 110 hernia repairs. Surg Radiol Anat. 2002;24:102–7.
4. Bower S, Moore BB, et al. Neuralgia after inguinal hernia repair. Am Surg. 1996;62:664–7.
5. Kumar S, Wilson R, et al. Chronic pain after laparoscopic and open mesh repair of groin hernia. Br J Surg. 2002;89:1476–9.
6. Ekberg O, Persson NH, et al. Longstanding groin pain in athletes. A multidisciplinary approach. Sports Med. 1988;6:56–61.
7. Bradshaw C, McCrory P. Obturator nerve entrapment. A cause of groin pain in athletes. Am J Sports Med. 1997;25:402–8.
8. Cohena S, Foster A. Pulsed radiofrequency as a treatment for groin pain and orchialgia. Urology. 2003;61:645.

9. Werner MU, Bischoff JM, et al. Pulsed radiofrequency in the treatment of persistent pain after inguinal herniotomy: a systematic review. Reg Anesth Pain Med. 2012;2012:340–3.
10. Aasvang E, Kehlet H. Surgical management of chronic pain after inguinal hernia repair. Br J Surg. 2005;92:795–801.
11. Lee C, Dellon L. Surgical management of groin pain of neural origin. J Am Coll Surg. 2000;191:137–42.
12. Loos MJ, Scheltinga MR, et al. Tailored neurectomy for treatment of postherniorrhaphy inguinal neuralgia. Surgery. 2010;147(2):275–81.
13. Zacest AC, Magill ST, et al. Long-term outcome following ilioinguinal neurectomy for chronic pain. J Neurosurg. 2010;112(4):784–9.
14. Alfieri S, Amid K, et al. International guidelines for prevention and management of post-operative chronic pain following inguinal hernia surgery. Hernia. 2011;15:239–49.
15. Chen DC, Hiatt JR, et al. Operative management of refractory neuropathic inguinodynia by a laparoscopic retroperitoneal approach. JAMA Surg. 2013;148(10):962–7.
16. Song JW, Wolf Jr JS, et al. Laparoscopic triple neurectomy for intractable groin pain: technical report of 3 cases. Neurosurgery. 2011;68(2):339–46.
17. Madura J, Madura II J, et al. Inguinal neurectomy for inguinal nerve entrapment: an experience with 100 patients. Am J Surg. 2005;189:283–7.
18. Fransen P. Epidural neurostimulation for chronic pain following abdominal hernia repair. Neurochirurgie. 2007;53(1):36–8.
19. Lepski G, Vahedi P, et al. Combined spinal cord and peripheral nerve field stimulation for persistent post-herniorrhaphy pain. Neuromodulation. 2013;16:84–9.
20. Mironer Y, Monroe TR. Spinal-peripheral neurostimulation (SPN) for bilateral postherniorrhaphy pain: a case report. Neuromodulation. 2013;16:603–6.
21. Rosendal F, Moir L, et al. Successful treatment of testicular pain with peripheral nerve stimulation of the cutaneous branch of the ilioinguinal and genital branch of the genitofemoral nerves. Neuromodulation. 2013;16(2):121–4.
22. Stinson LW, Roderer GT, et al. Peripheral subcutaneous electrostimulation for control of intractable post-operative inguinal pain: a case report series. Neuromodulation. 2001;4:99–104.
23. Yakolev A, Al Tamimi M, et al. Spinal cord stimulation as an alternative treatment for chronic post-herniorraphy pain. Neuromodulation. 2010;13(4):288–91.

Part 7

Rehabilitation

18 Conservative Rehabilitation Treatment in Groin Pain Syndrome

Ilaria Maria Marchetti, Cristiano Sconza, and Stefano Respizzi

18.1 Introduction

Evaluation and treatment of groin pain syndrome in athletes is challenging since different pathologies may cause similar symptoms and many systems can refer pain to the groin. All the anatomic structures in the hip and groin region, like muscles, tendons, nerves, ligaments and joints, interact and depend on each other [1]. In addition to a complex anatomy, we have a varied range of pathologies originating from multiple aetiological processes that can cause activation of pain fibres in the inguinal region [2]. The origin of pain can be simple as a muscle injury, caused by a direct trauma or an adductor strain, or more complex as a "sport hernia". The groin pain could also be due to a referred pain that may come from different anatomical regions like the spine, the sacroiliac joint, the testes or the urinary tract. Other possible causes could be stress fractures affecting the femoral neck or the inferior pubic ramus, osteitis pubis or hip pathologies like femoro-acetabular impingement (FAI), acetabular labral tears, hip instability or osteoarthritis [3]. The possibility of coexisting injuries further complicates the management of these patients: if more than one affliction can cause groin pain, it is difficult to establish which is the main contributor and which is secondary. It has been shown that treating more than one condition at the same time can lead to better clinical results: for example, when an athlete has a surgical sports hernia repair and a hip arthroscopy at the same time, the success rate is much better than performing surgery on just one pathology [4]. These observations suggest that groin pain has a complex pathogenesis and is connected to various diseases, each having its role in determining the risk of injury.

The conservative treatment of patients with groin pain could be even more complicated. When the athlete first come to the rehabilitation specialist observation, in a lot of cases he has already received a non-specific diagnosis of groin pain, and he has usually tried to take a prolonged period of rest and different types of treatments. These confounding factors and the insidious onset of the pathology can make the diagnostic process really challenging. In addition, during the rehabilitation programme, the patient may experience periods of improvement interspersed with episodic pain exacerbations, making the long-term success sometimes difficult. As regards the

I.M. Marchetti (✉)
Sport & Exrcise Medicine, University of Milan, Milan, Italy
e-mail: ilaria04071987@gmail.com

C. Sconza
Department of Rehabilitation and Functional Recovery, Istituto Clinico Humanitas, via Manzoni 56, Rozzano (MIlan), Italy
e-mail: cristiano.sconza@gmail.com

S. Respizzi
Department of Rehabilitation and Functional Recovery, Istituto Clinico Humanitas, via Manzoni 56, Rozzano (MIlan), Italy
e-mail: stefano.respizzi@humanitas.it

R. Zini et al. (eds.), *Groin Pain Syndrome*, DOI 10.1007/978-3-319-41624-3_18

current literature, the available evidence provides some useful starting points, but it is insufficient to give universal rehabilitation guidelines [5].

For all these reasons, we think that it's primarily important to start from a proper evaluation for identifying a correct and complete diagnosis and correlate it with the mechanism and biomechanical aspects of the injury. Each patient has a distinct injury history and unique clinical findings, thus only understanding the specific dysfunction, we can assess an effective customized rehabilitation programme.

The aim of a conservative rehabilitation project is to treat not only the current injury but also all the underlying risk factors. For example, we must consider that sports involving sudden changes of direction, acceleration, deceleration, sliding tackles and kicking (e.g. football, hockey, basket) can lead to muscular overload and imbalance. The overload of pubic symphysis and insertional tendons could be induced by strength imbalance between hip adductor muscles and rectus abdominis [6]. Literature suggests that coexisting postural impairments like reduced flexibility of the posterior chain muscles, lumbar hyperlordosis with pelvic anteversion and dissymmetry of the lower limbs often lead to a pelvic complex instability that can predispose the athlete to groin-related injuries [6].

When a patient comes to our observation complaining of long-term groin pain, a meticulous postural examination has a crucial importance in order to evaluate even minimal dysfunctions that, in combination with muscular imbalance, can lead to groin pain syndrome especially in professional athletes. Initially the patient is examined in an orthostatic position, to assess the symmetry of the pelvis and the shoulders. It is important to observe the plantar support with the assistance of the podoscope and the structure of the hindfoot and forefoot. Mobility on all planes of the lumbosacral rachis should be examined as well as the presence or absence of scoliosis or scoliotic posture. Lateral examination of spinal curvatures, rotation of the pelvis and posture of hip and knees should be done. A typical report of the adductor syndrome is the lumbar hyperlordosis with pelvis anteversion. In case of suspected inguinal hernia or sports hernia, it is helpful to evaluate the testicles and the inguinal canal with the Valsalva manoeuvre. Painful points like tendon insertions of adductor, recto-abdominal and iliopsoas muscles, pubic symphysis and iliac spines should be evaluated. Pain can also be reproduced by isometric contraction of abdominal muscles, iliopsoas, rectus femoris and adductor muscles or by passive stretching of the anterior chain (Thomas test) and the posterior chain. A specific examination of the hip joint should not be forgotten: the articular mobility of the hip on all planes should be assessed with the patient lying in supine position. Reduced joint ROM (range of motion), especially in hip internal rotation, flexion and adduction (FADDIR test), is a common finding in people with FAI and early osteoarthritis. Impaired posture of the spine and pelvis can also predispose to FAI, like the swayback posture (see the chapter below). Evaluation of hip muscle strength and length can provide further information to confirm the diagnosis and guide the treatment [6]. Finally, a peripheral neurological examination should be conducted in order to exclude radiculopathy or other neurological diseases.

In some cases the rehabilitation specialist needs to perform imaging techniques, like plain a radiograph, an ultrasound scan or a magnetic resonance (MRI). The can be useful to confirm the diagnosis or to exclude different causes of groin pain, including hip osteoarthritis or fractures [7].

In order to illustrate the concepts that underlie the preparation of a rehabilitation programme, we are going to describe the most frequent pathologies that cause groin pain, referring to current literature. Remember once again that, despite the strategic importance of a correct diagnosis, only treatment of specific clinical findings and coexisting risk factors can guarantee the clinical success.

18.2 Conservative Treatment in Tendon and Muscle Pathology

Insertional tendinopathy of the hip adductor muscles and rectus abdominis is a really frequent cause of groin pain, with an incidence of 2.5–3% among elite athletes [7]. The most frequent involved activities are soccer, basketball, hockey, rugby and long-distance running. It is considered

Table 18.1 Example of exercise therapy protocol that could be used for the conservative rehabilitation treatment of adductor-related groin pain

First phase (first 2 weeks)	Second phase (from 3rd week)	Third phase (from 6th week to return to run)
Static adduction against ball placed between feet lying supine	Leg abduction and adduction exercises performed in side lying	Slow jogging • Slow running at easy pace • Can progress if no pain
Static adduction against ball placed between knees when lying supine	Low-back extension exercises prone over end of couch	Straight sprints • 100 m sprints at increasing speed and repetition • Can progress if no pain
Abdominal sit-up both in straightforward direction and in oblique direction	One-leg weight pulling abduction/adduction standing	Cutting • Sport-specific sprints involving change of directions
Combined abdominal sit-up and hip flexion, starting from supine position and with ball between knees	Abdominal sit-up both in straightforward direction and in oblique direction	
Balance training on wobble board	One-leg coordination exercise with flexing and extending knee and swinging arms in same rhythm	
One-foot exercises on sliding board, with parallel feet as well as with 90° angle between feet	Training in sideways motion on a mini-skateboard	
	Balance training on wobble board and skating movements on sliding board	

The protocol is divided in three chronological progressive phases. The athletes start to perform the exercises of the first phase; after indicatively 2 weeks, they can pass to the second phase. At 6 weeks they can start the return to running programme [10]

an overuse disease, but it can also occur as a result of a previous injury like a groin strain. It consists of inflammation (in the acute phase) and consequently degeneration (in the chronic phase) of tendons directly at their insertion on the ischio-pubis bone, and it can involve the common adductors tendon and/or the rectus abdominis tendon [8].

The conservative treatment consists of a multidisciplinary approach depending on the clinical phase of the pathology and progresses through functional steps.

It has been shown that the addition of an active physical training is more effective than the use of passive treatments alone and that a multimodality approach, including manual therapy techniques, can provide a quicker return to sport in the athlete (Table 18.1) [9, 10]. Active training of strength and coordination of muscles acting on the pelvis, in particular the adductors, is very effective in the treatment of insertional tendinopathy and groin pain in general [1].

The first step of the conservative treatment is focused on inflammation and pain control. We can use anti-inflammatory drugs, instrumental and manual therapy. Pharmacotherapy consists of systemic administration or local injection of NSAIDs, corticosteroids and, recently, platelet-derived growth factor (PDGF) [6].

Heat therapy with resistive to capacitive system, laser therapy and electrical therapy can be really useful in order to decrease local swelling and inflammation and to reduce pain acting on nociceptor's membrane and the nervous fibres by the modulation of the painful stimulus and its transmission. Extracorporeal shockwave therapy can also facilitate tendon enthesis healing and regeneration due to the stimulation of nitric oxide and growth factor production and stem cell proliferation, migration and differentiation. Physical therapy is recommended as the first-line treatment after a period of rest or restricted activity [11]. In the acute phase, we can use postural balance techniques through global and site-specific stretching, mechanical and proprioceptive orthotic insoles and global postural re-education. Improving the hip ROM, specifically internal and external rotation, has been proposed as a possible method to reduce stress across the pubic symphysis and the surrounding structures [12]. Muscular imbalance between abdominal and hip muscles may contribute to mechanical overload and loss of functional stability of the groin [1]. Hip adductors are essential stabilizers of the pelvis together with gluteus, hamstrings and abdominal muscles in sport activities like running, pivoting and kicking [13]. Thus, it may be possible that improving control and strengthening of these muscles [14] may restore the function of the groin region [5]. Decontracting massotherapy is useful to relax and stretch adductors and abdominal muscles. A correct application of neuromuscular taping can detent tendon insertions, promote muscle relaxation and protect the muscle-tendon unit from overstretching. In the early stages, active exercise therapy involves isometric strengthening of abdominal and adductor muscles in the gym or in the warm water of a therapeutic swimming pool. In the subacute phase, muscle strengthening is increased by introduction of concentric and eccentric exercises and cardiovascular reconditioning. Core stability exercises should be introduced as soon as possible and consist of the contextual and synergic activation of abdomen, adductor and lumbar muscles. Finally, running is gradually started at first on a treadmill, proceeding with free aerobic run at increasing speed. The last rehabilitation step will be focused on the return to sport [7]. Learning preventive postural, eccentric and plyometric exercises is important during and after the return-to-sport phase in order to maintain a good stretch of the posterior and adductor chain, as well as a global balance between agonist and antagonist muscle groups [6].

18.3 Conservative Management of Sport Hernia

Sport hernia is defined as a weakness of the posterior inguinal wall without a clinically palpable hernia, which results from injury of muscular and/or fascial attachment to the anterior pubis bone [15]. Tears associated with an athletic pubalgia may involve the transversalis fascia at the posterior inguinal wall, the insertion of the distal rectus abdominis, the conjoined tendon and the external oblique aponeurosis [15]. The hallmark symptom is a severe lower abdominal, pubic or groin pain with exertion that improves with rest but typically returns upon resumption of sport-related activities. Pain is usually deeper, more proximal and more intense than an adductor or rectus abdominis strain. It is common to find a point tenderness near the lower abdominal insertion at the pubic tubercle that can involve the adductor longus tendon origin as well.

Rehabilitation is usually the first option in treating a nonathlete patient with a diagnosis of sport hernia because evidence suggests poor surgical outcomes in these cases [16]. Moreover most physicians agree that a conservative treatment should be in many cases the first option even in athletes [15]. The current literature shows significant improvement after 6–8 weeks of physical therapy intervention [17]. The duration of the rehabilitation process depends on the nature of the injury, the level of preinjury performance of the athlete and the length of time before the athlete is expected to return to play [15]. The diagnosis of sport hernia usually implicates several weeks of rest and a gradually return to sports activities, depending on the single patient's injury history.

First of all, the conservative treatment consists in pain control and reduction of swelling and inflammatory reaction. The patient should be able to move and conduct daily activities with little to no pain present. We can use oral anti-inflammatory drugs, physical agents or local injections in order to accelerate the reduction of symptoms. The rehabilitation treatment is mainly based on the association of exercise therapy and manual therapy. It has been shown that the use of these two treatments in combination allows achieving better results than using exercise therapy alone, with a faster return to sport [10]. Manual therapy works by improving soft tissue and fascia restrictions, reducing the viscoelastic stiffness in the adductor

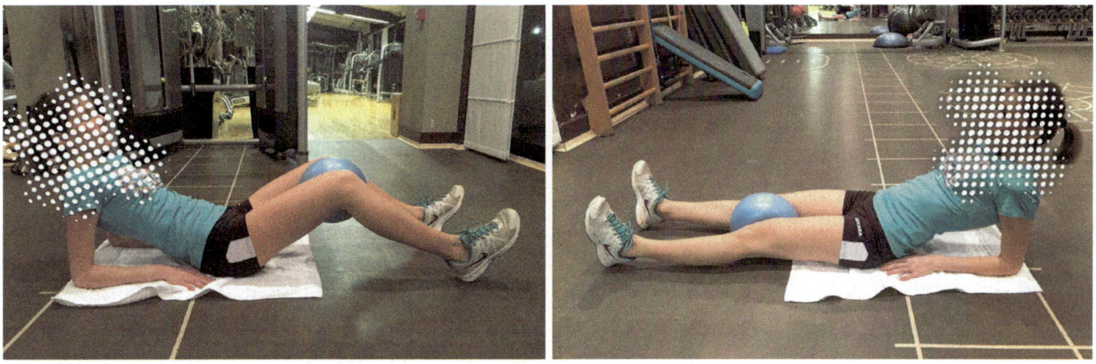

Fig. 18.1 Adductor squeeze test: pain provocation by isometric hip adduction. It can be performed at 0° (on the *right*), 30° (on the *left*), 45° and 90° hip flexion. It is possible to quantify the adductor strength during this test through pressure values measurement (mmHg), using a simple sphygmomanometer pre-inflated to 10 mmHg [18]

muscles and mechanical loads on the pubic bone. It also seems that manual therapy can influence core stability: the manipulation of the sacroiliac joint can enhance abdominal muscles function, improving the feedforward action of transversus abdominis muscle. Those techniques include soft tissue manipulation of the lumbar and hip regions, mobilization of the pelvis, sacroiliac and hip joints, neuromuscular re-education and manual stretching [10]. The main part of the conservative treatment is based on a customized and progressive exercise therapy programme that aims to minimize pre-existing risk factors and compensatory strategies, implement core stabilization and maintain good motor control around the pelvis. Safe progression through the various stages of the rehabilitation programme requires strict monitoring using objective tests and measures like the adductor squeeze test (Fig. 18.1) and the hip range of motion.

All interventions address specific objective impairments found during the physical examination, including muscle length deficits, strength imbalances and joint hypo-mobility. In fact these impairments may have led to, or been the result of, the sports hernia injury.

Clinical assessment of core stability, and identification of hip muscular compensation and imbalance are crucial. Treatment should target strengthening and neuromuscular re-education regarding timing and recruitment patterns during functional motion. In the early phase, the patient can perform a sequence of submaximal isometric contraction of hip abductors, adductors, flexors and extensors. Core stability exercises begin with static contraction of the deep core muscles and an initial emphasis on correct activation. These exercises are progressed by adding single-leg activities in stable positions and, subsequently, in unstable positions, incorporating external perturbation in order to improve proprioception and kinesthetic awareness [19]. Functional strength training includes a pattern of bilateral squat to single-leg squat and lunge, adding direction to lunge in the final stages. Dynamic stabilization of the pelvic ring is the central goal [20]. Active stretching of the spine and lower extremities to ensure the preservation of flexibility and full range of motion should be added. Joint range of motion of the hips and lumbar spine is expected to be optimal prior the return to sport.

The final step of the rehabilitation programme of all athletes with groin pain syndrome is the return to sport phase. Once completed this step, the athlete is expected to return to full competitive sport. This phase should begin with a gradual return to running; the player may start with a short run every 2 days improving gradually under supervision. Clinical criteria to start a safe return to running should be pain-free adductor squeeze test, minimal adductor guarding (i.e. spasm of the adductor muscles on passive hip abduction), pain-free pubic symphysis shear test into extension (feel for hypermobile pubic symphysis while shearing vertically through anterior superior iliac spines) and pain-free fast walking [20]. Running is progressed into straight-line speed, changing direction and then accelerative drills. Finally it can be introduced the sport-specific training on the field. Physicians and trainers typically allow the return to play once the athlete has no pain with sport-specific activities like sprinting and cutting and he can complete an abdominal curl-up and bilateral straight-leg raise/hip flexion without symptoms.

18.4 Conservative Management of Femoroacetabular Impingement

In recent times the hip joint has been recognized as a significant cause of hip and groin pain in the athletic population. Groin pain is frequently reported in patients attending for hip arthroscopy, evidenced by 92% of patients with labral tears [21]. As previously described, hip pathologies often coexist with other groin-related diseases including adductor strain and sports hernia, which can make definitive diagnosis and management difficult and often multifactorial. Recent studies have found that 94% of athletes with adductor-related groin pain have radiological signs of FAI [22]. FAI is not considered to be a hip pathology; it is recognised as a morphological variant seen in approximately 20% of general population that may increase the risk of intra-articular hip pathology including labral tears and chondropathy, contributing to the development of groin pain. When the hip joint is placed into a position of impingement in a repetitive fashion during sport activities, microtraumas may occur due to the friction of labrum between the hip bony components, causing labral tears [23]. Moreover the presence of FAI and labral pathology may lead to an increased risk of chondropathy and early hip osteoarthritis [24].

The published literature suggests that conservative treatment is a reasonable first choice for patient with FAI [25]. However, no studies have compared the effectiveness of hip arthroscopy to conservative management. Depending on multiple factors such as the clinical presentation, presence of comorbidity and time in season, the current evidence suggest to attempt a period of at least 6–8 weeks of conservative treatment prior to an eventual surgical treatment. This is a very general recommendation that must be individualized to the patient [26].

Current strategies for the conservative management are mainly based on the clinical findings of physical impairments in people with FAI. Rehabilitation programme focuses on modifying adverse hip joint forces created by abnormal hip morphology and pathology. This can be achieved by three different steps: improving hip muscles function, modifying hip external loads (such as the volume and intensity of physical activity) and providing an appropriate education for activities of daily living predisposing to FAI [27].

Hip muscle strength and length and hip joint ROM are usually limited in patients with FAI. Impaired biomechanics of the hip and pelvis is present, which may partially explain the association with groin pain. Individuals with FAI usually present a swayback posture that consists of loss of normal lumbar curve with the pelvis swayed forward in front of the centre of gravity and tilted posteriorly and the hip joint extended. In this type of posture, the external oblique muscles and the iliopsoas are excessively stressed, while the hip flexors such as rectus femoris and tensor fascia latae are short. The line of gravity moves posteriorly to the hip joint, resulting in disuse atrophy of the gluteal muscles. The stiffness of the hip extensors and posterior joint structures and the excessive flexibility of anterior hip joint structures result in a path of least resistance to anterior glide [26]. Hip muscle strength impairment contributes to modify intra-articular hip loads: a typical finding in these patients is a weakness of gluteus medius and maximus and iliopsoas, while the tensor fascia latae is short and dominant. Limited ROM and pain during hip internal rotation, flexion and adduction (FADDIR or anterior impingement test) is a classical physical examination finding in people with symptomatic FAI. Probably maximum hip flexion and extension increases load on the anterior and superior joint regions, which is the anatomical site

Fig. 18.2 Example of exercises aimed at restoring trunk and deep hip stabilizer muscle function

associated with the majority of FAI lesions. Another common finding in these patients is a positive FABER test: during flexion, abduction and external rotation of the hip joint, the vertical distance of the knee to the examination table is increased on the symptomatic hip in comparison to the unaffected side [28]. Muscle function and ROM recovery with suitable rehabilitation strategies have the potential to modify hip joint loads and mitigate the progression of hip and groin pain. Improvement hip muscle strength appears to be an important treatment goal for this purpose. It is fundamental to focus at first on the primary deep stabilizer of the hip (iliopsoas, gluteus maximus, gluteus medius, quadratus femoris, obturator internus, inferior and superior gemelli, adductor brevis and pectineus) which are thought to provide a posterior, medial and inferior force on the femur, ensuring the head of the femur located in a position that minimize stress on the anterosuperior acetabular labrum and rim (Fig. 18.2).

The goal is to improve posterior glide of the femur, decreasing the mechanical contact between the acetabular edge and the femoral neck. This can be achieved in part with muscular strengthening and in part using joint passive mobilization. The patient's hip is flexed, mild abducted and internally rotated with the foot fixed on the table; the clinician provides a posterior-lateral force directed longitudinally through the femur if tolerated. Joint mobilization is followed by self-mobilization and active exercises to maintain mobility.

Once adequate control of the deep hip stabilizers has been obtained, a staged hip-strengthening programme can be undertaken.

As regards the external loads, there is minimal evidence examining the influence of type and volume of physical activity on hip joint loads. However probably the intensity of the activity and the hip ROM required may increase anterior hip joint forces, like when the hip is extended beyond 10° of ROM. This force increases when weakness in the gluteal muscles or iliopsoas is present [29]. Moreover, there is evidence that a great amount of high impact physical activity is associated with a raised risk of hip osteoarthritis.

It appears that hip joint loads can be modulated by type and amount of physical activity; therefore rehabilitation programmes should include strategies to address the volume and intensity of activity undertaken.

Another key target in these patients is to modify activity of daily living predisposing to FAI. The safe range of motion to avoid FAI is between maximum internal and external rotation, avoiding hip internal rotation associated with flexion and adduction. The patient has to be instructed to adapt to this safe range of movement in activities of daily living with minimal friction. Stretching exercises can help to improve hip external rotation and abduction and to avoid the "W" sitting position, with the hip in abduction and internal rotation. The patient should avoid sitting continuously of a long time with the spine fully straight and the hip in flexion; leaning back every 5 min is encouraged as it can decrease stress on the acetabular labrum. Cycling and running on a treadmill or narrow straight trails are not recommended to prevent hip flexion and internal rotation. Using primarily this behavioural approach, encouraging results have been achieved in young sportive patients, without limiting the level of physical activity [30].

Before the return to sport, the athlete with FAI may benefit from a running training with increasing step rate in order to avoid any improper running technique. When satisfactory muscular control is regained, we focus on running form, and the athlete is ready to start a specific graded programme. The runner should begin on softer surface, such as grass or track. As form and confidence improves without pain, speed may be increased gradually.

References

1. Holmich P, Uhrskou P, Ulnits L, et al. Effectiveness of active physical training as treatment for long-standing adductor-related groin pain in athletes: randomised trial. Lancet. 1999;353:439–43.
2. Jacob BP, Chen DC, Ramshaw B, et al. The SAGES manual of groin pain. Cham (ZG): Springer; 2016.
3. Gilmore J. Groin pain in the soccer athlete: fact, fiction, and treatment. Clin Sports Med. 1998;17:787–93.
4. Larson CM, Pierce BR, Giveans MR. Treatment of athletes with symptomatic intra-articular hip pathology and athletic pubalgia/sports hernia: a case series. Arthroscopy. 2011;27:768–75.
5. Almeida MO, Silva BNG, Andriolo RB, et al (2013). Conservative interventions for treating exercise-related musculotendinous, ligamentous and osseous groin pain. Cochrane Database Syst Rev. (6):CD009565. doi:10.1002/14651858.CD009565.pub2.
6. Valent A, Frizziero A, Bressan S, et al. Insertional tendinopathy of the adductors and rectus abdominis in athletes: a review. Muscles Ligaments Tendon J. 2012;2:142–8.
7. Volpi P. Football traumatology, new trends. 2nd ed. Cham (ZG): Springer; 2015.
8. Schilders E, Bismil Q, Robinson P, et al. Adductor-related groin pain in competitive athletes. Role of adductor enthesis, magnetic resonance imaging, and entheseal pubic cleft injections. J Bone Joint Surg Am. 2007;89:2173–8.
9. Serner A, van Eijck CH, Beumer BR, et al. Study quality on groin injury management remains low: a systematic review on treatment of groin pain in athletes. Br J Sports Med. 2015;49:813.
10. Weir A, Jansen JACG, van de Port IGL, et al. Manual or exercise therapy for long-standing adductor-related groin pain: a randomised controlled clinical trial. Man Ther. 2011;16:148–54.
11. Jansen JACG, Mens JMA, Backx FJG, et al. Diagnostics in athletes with long-standing groin pain. Scand J Med Sci Sports. 2008;18:679–90.
12. Verrall GM, Slavotinek JP, Fon GT, et al. Outcome of conservative management of athletic chronic groin injury diagnosed as pubic stress injury. Am J Sports Med. 2007;35:467–74.
13. Biedert RM, Warnke K, Meyer S. Symphysis syndrome in athletes: surgical treatment for chronic lower abdominal, groin, and adductor pain in athletes. Clin J Sport Med. 2003;13:278–84.
14. Cowan SM, Schache AG, Brukner P, et al. Delayed onset of transverse abdominus in long-standing groin pain. Med Sci Sports Exerc. 2003;36:2040–5.
15. Kachingwe AF, Grech S. Proposed algorithm for the management of athletes with athletic pubalgia (sports hernia): a case series. J Ortho Phys Ther. 2008;38:768–81.
16. Meyers WC, Lanfranco A, Castellanos A. Surgical management of chronic lower abdominal and groin pain in high-performance athletes. Curr Sports Med Rep. 2002;1:301–5.
17. Woodward JS, Parker A, MacDonald RM. Non-surgical treatment of a professional hockey-player with the signs and symptoms of sports hernia: a case report. Int J Sports Phys Ther. 2012;7:85–100.
18. Delahunt E, Kennelly C, McEntee BL, et al. The thigh adductor squeeze test: 45 of hip flexion as the optimal test position for eliciting adductor muscle activity and maximum pressure values. Man Ther. 2011;16:476–80.
19. Tyler TF, Silvers HJ, Gerhardt MB, et al. Groin injuries in sports medicine. Sportshealth. 2010;2:231–6.
20. Garvey JFW, Read JW, Turner A. Sportsman hernia: what can we do? Hernia. 2010;14:17–25.
21. Burnett RSJ, Della Rocca GJ, Prather H, et al. Clinical presentation of patients with tears of the acetabular labrum. J Bone Joint Surg Am. 2006;88:1448–57.
22. Weir A, de Vos RJ, Moen M, Hölmich P, et al. Prevalence of radiological signs of femoroacetabular impingement in patients presenting with long-standing adductor-related groin pain. Br J Sports Med. 2011;45:6–9.
23. Ito K, Minka MA, Leunig M, et al. Femoroacetabular impingement and the cam-effect. A MRI-based quantitative anatomical study of the femoral head-neck offset. J Bone Joint Surg Br. 2001;83:171–6.
24. Pollard TCB. A perspective on femoroacetabular impingement. Skeletal Radiol. 2011;40:815–8.
25. Wall PD, Fernandez M, Griffin DR, Foster NE. Nonoperative treatment for femoroacetabular impingement: a systematic review of the literature. PM R. 2013;5:418–26.
26. Loudon JK, Reiman MP. Conservative management of femoroacetabular impingement (FAI) in the long distance runner. Phys Ther Sport. 2014;15:82–9.
27. Kemp J, Crossley K. Conservative management of femoroacetabular impingement. A case study and

rationale for treatment. Aspetar Sport Med J. 294–299. www.aspetar.com/journal.
28. MacIntyre K, Gomes B, MacKenzie S, et al. Conservative management of an elite ice hockey goaltender with femoroacetabular impingement (FAI): a case report. J Can Chiropr Assoc. 2015;59:398–409.
29. Lewis CL, Sahrmann SA, Moran DW. Anterior hip joint force increases with hip extension, decreased gluteal force, or decreased iliopsoas force. J Biomech. 2007;40:3725–31.
30. Emara K, Samir W, Motasem ELH, et al. Conservative treatment for mild femoroacetabular impingement. J Orthop Surg (Hong Kong). 2011;19:41–5.

Postsurgical Rehabilitation: FAI

19

Maria Teresa Pereira Ruiz

19.1 Introduction

Rehabilitation following hip arthroscopy has long been recognized as an integral component of the clinical outcome of the procedures [1].

Over the past 10 years, the number of hip arthroscopies has increased in an important way [2]. Based on the increase of this type of surgery, its complexity, and the diversity of procedures performed, the creation of appropriate rehabilitation protocols to adequately satisfy the needs of the different techniques.

The etiology of femoroacetabular impingement (FAI) is multifactorial, and it can be post-traumatic, it can be a sequela of a pediatric disease, or it can be idiopathic [3]. In this way, before speaking of a rehabilitation program after arthroscopic surgery, it is essential to carefully study patient history. It becomes very important for the success of the rehabilitation process to examine any muscle imbalances and postural changes that we need to try to correct during the rehabilitation program. Another fundamental point is the collaboration between the surgeon and the physiotherapy team that treats him. In this article we will try to analyze the various rehabilitation programs that there are in literature and propose the one we use based on our clinical experience.

19.2 Current Concepts of Rehabilitation

In recent years there are several works in literature that propose different protocols of rehabilitation after hip arthroscopy. At this point, works that carry out a systematic review concluded that "the current literature of hip arthroscopy rehabilitation lacks high-quality evidence of support to specific protocol" [4].

For a proper setting of the rehabilitation protocol, we first need to know specifically which type of surgery was performed. The type of operation together with the type of patient will allow us to set the proper protocol for that patient. Currently we can summarize the surgical techniques in labral debridement, femoral osteochondroplasty, pincer acetabuloplasty, teres ligament debridement, acetabuloplasty rin trimming, labral repair, microfracture of the femoral head and/or acetabulum, loose body removal, and iliopsoas release [4]. We can also find several techniques together in one operation. This makes us understand that the information and protocols described are to be used as a guide that permits us to tailor it to suit each patient.

M.T.P. Ruiz
Montallegro Clinic, Genoa, Italy

Isokinetic, Milan, Italy
e-mail: mtpereiraz@gmail.com

R. Zini et al. (eds.), *Groin Pain Syndrome*, DOI 10.1007/978-3-319-41624-3_19

19.3 Weight-Bearing Status and Range of Motion

The majority of the authors propose weight bearing as tolerated in cases of labral debridement, loose body removal, and osteoarthritis debridement. In cases of labral repair, femoral osteochondroplasty, and pincer acetabuloplasty, we recommend a partial weight bearing or slight weight bearing for 3–4 weeks. In case of microfractures, the literature proposes protected bearing time for up to 6 weeks [3].

With regard to the limitations in range of motion (ROM), there are differences in the literature. All guidelines suggest initiating early, protected ROM. The utilization of gentle ROM circumduction for the hip joint has been suggested early in the rehabilitation process [3, 5]. Some authors propose a limited flexion up to 90° for 2 weeks (no limit extension, rotation, or abduction) for labral debridement, flexion limited up to 90° for 2 weeks, extension to 10° for 2 weeks, abduction 25° for 2 weeks, and gentle rotation for 2 weeks for labral repair [6]. With a labral repair, the location and side of the tear should be held in consideration. Most of these tears are located in the anterior or anterosuperior portion of the labrum; movements that stress this area should be avoided [5, 7]. Others suggest a gradual recovery of the ROM without limits, to prevent adhesions [8]. In the case of tissue release procedures, avoidance of early initiation of long lever hip flexion (iliopsoas release) or abduction (iliotibial release) is suggested [9].

19.4 Rehabilitation Phases

19.4.1 Phase I

It is a protection period, with a duration of 0–4 weeks in simple cases with little osseous degenerative pathology and with good structural stability and neuromuscular control that will rapidly progress (uncomplicated) and 0–6 weeks in individuals who have a more complex combination of osseous abnormalities and poor neuromuscular control (complicated). The goal is to reduce lower limb edema, gently progress hip range of motion, and regain normal neuromuscular firing patterns of the pelvis and hip.

We recommend that, during the first 2 weeks at home, the program must be performed two times a day. This program involves isometric exercises for the transversus abdominis, quadriceps, and gluteus, exercises for the ankle pump, and gentle hip circumduction. The progression depends on the patient's tolerance and should not be overly aggressive. One key component of the home program is local stabilization of the hip joint by retraining and strengthening the deep hip rotator muscles. This includes the deep musculature quadratus femoris, the gemellus superior and gemellus inferior, and the obturator internus. It has been suggested that they may, in the end, provide control of the hip joint stability, acting as "rotator cuff" of the hip joint [10, 11] (Fig. 19.1).

The aquatic program usually commences 3 weeks after surgery. This will initially consist of walking in the pool and using a stationary bike and cross-trainer. The use of hydrotherapy in gait retraining and weight bearing is very effective

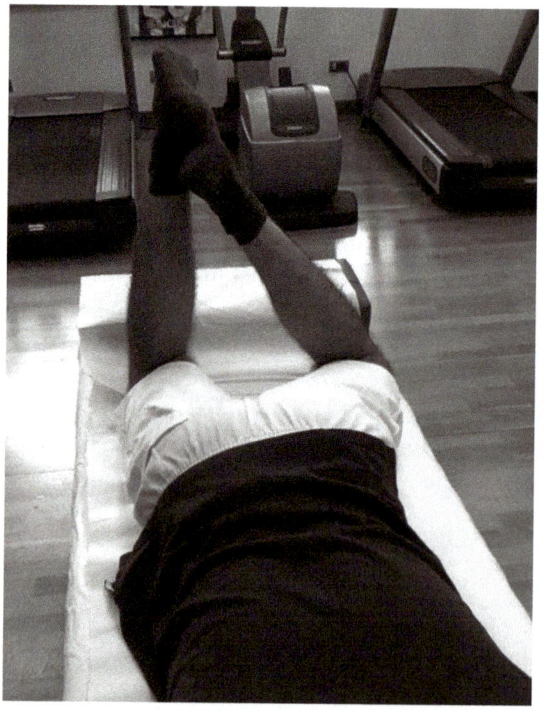

Fig. 19.1 Deep hip rotators activation facedown

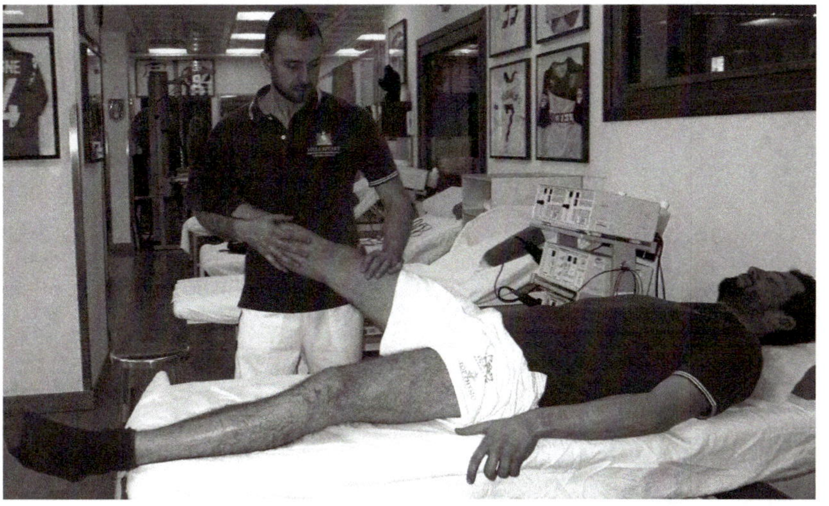

Fig. 19.2 Soft tissue mobilization and massage

once the incisions have healed. Lastly, proprioceptive exercises in bilateral stance are started as soon as weight-bearing restrictions are lifted.

After 2 weeks, the patient can begin working daily without resistance on a stationary bike. Soft tissue mobilization and massage focus on the adductor muscle group, which tends to quickly develop tone (Fig. 19.2). Clinically, it appears that while other pelvic and hip stabilizers are inhibited on account of pain or neuromuscular dysfunction, the adductors are often the first muscle group that compensate. Following hip arthroscopy, the psoas muscle is often inhibited. Clinically, the tensor fascia lata and the rectus femoris tend to compensate for the lack of the function of the psoas and become overused and irritated during the postoperative course. These muscles, along with the gluteus, benefit from massage to reduce tone throughout the rehabilitation process [3].

Criteria transition to Phase II: walking with full loads; absent or minimal pain with exercises of Phase I; proper muscle activation during the activity; no inflammation at the hip flexor.

19.4.2 Phase II

With a duration of 4–6 weeks for uncomplicated forms and 6–12 weeks for complicated forms, the goal of the second phase is for the patients to achieve independence in daily activities with little or no discomfort. The focus of this phase is to continue progressing ROM (pain-free) and soft tissue flexibility while beginning to transition the emphasis to strengthening. We continue with hip circumduction and deep massage and mobilization. It's very important to reeducate and address the psoas muscle imbalance. Edelstein suggests it is most effective to reeducate the psoas from the trunk down versus the leg up. The results have thus far been successful, in that the patients regain full hip flexion strength without developing tendonitis of the psoas or the secondary hip flexors. Progressive eccentric exercises for the psoas should be introduced. Another important point is the gluteal function, and this may be accomplished by having the patient lie prone and reeducate transversus firing, then gluteal firing, followed by a small hip extension motion being careful to not permit the activity to nullify the core stabilizers as shown by lumbar extension or pelvic rocking. Anterior (Fig. 19.3) and posterior capsular stretching (Fig. 19.4) and kneeling hip flexor stretch as tolerated. Phase II is the time to progress from closed chain bilateral dynamic stability exercises to unilateral exercises. Examples include forward step downs, three point stepping with elastic band, windmill, and lawnmower (Fig. 19.5).

Other exercises are core and hip and pelvis strengthening, adding resistance to the stationary

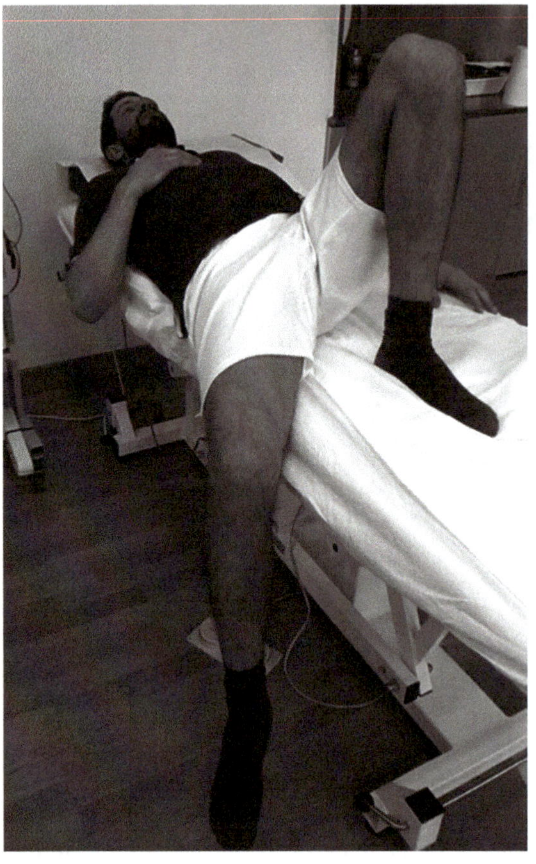

Fig. 19.3 Anterior capsular stretching

bike, for athletes progressive cardiovascular fitness and upper extremity.

Criteria transition to Phase III: Ambulation with full load in the absence of pain; complete recovery of ROM; recovery of hip flexion strength >60% compared to the contralateral limb; abduction, adduction, external rotation, internal rotation >70% compared to the contralateral limb. Precautions: No ballistic stretching; no running on the treadmill.

19.4.3 Phase III

With a duration of 8–12 weeks for uncomplicated forms and 12–20 weeks for complicated forms, the goal is to become asymptomatic recreationally. Strengthening exercises should now incorporate multi-planar movement involving multiple muscle groups. Manual therapy should be performed as needed. Flexibility and passive ROM exercises should become slightly more aggressive if limitations persist. The patient will need periodic verbal and tactile cueing for both trunk stabilization and body position awareness, especially when they begin to tire. The proprioceptive retraining program should reestablish neuromotor control and is an important part of

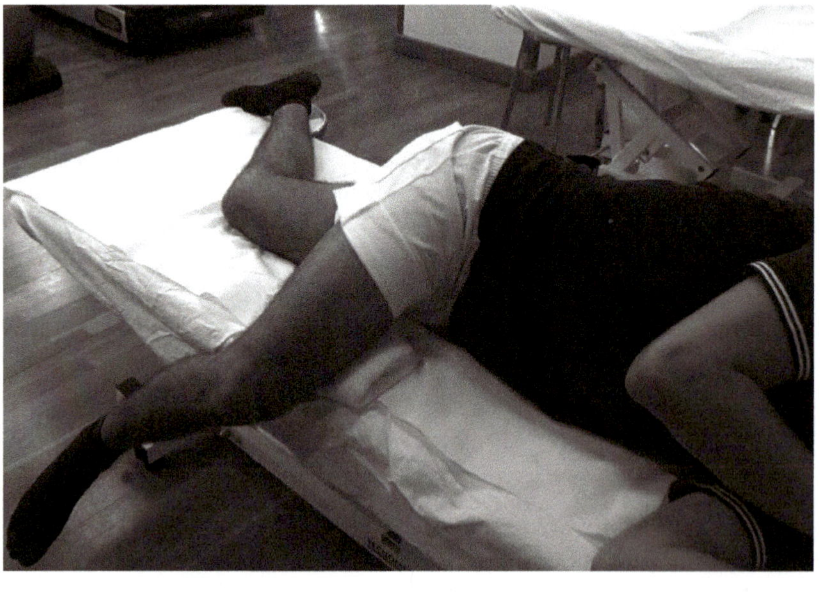

Fig. 19.4 Posterior capsular stretching

Fig. 19.5 Elastic band in three point

this phase. Depending on patient tolerance, the exercise program may progress from slow to fast, simple to complex, stable to unstable, low to high force, and general to specific [12, 13]. For athletes, it is important to increase the volume and intensity of aerobic activities. Core exercises initiated during this phase may include kneeling cable crossovers, T-position side supports, T-position eye of the needle, side support rowing, side support hip abduction, front leg support, and raise kneeling chops/lifts, for example. Edelstein et al. suggest that prior to commencing running, the person should be capable of asymptomatically performing the "Triple 10 Rep." This includes ten front step downs without kinetic collapse, ten single-leg squats without kinetic collapse, and ten side-lying leg raises against resistance with manual muscle with testing score 4/5 for all repetitions without compensation [3].

Criteria transition to Phase IV: Recovery of hip flexion strength >70% compared to the contralateral; abduction, adduction, external rotation, internal rotation >80% compared to the contralateral; ability to perform agility drills (sports cords forward/back run, multi-planar lunges).

19.4.4 Phase IV

The duration for uncomplicated forms is 12–16 weeks and 20–28 weeks for complicated forms. The primary objective of this phase is a safe and effective competition or return to previous activity level. Phase IV generally requires pain-free full-motion, strength, without any subjective or objective deficit during training. In literature, there is only one study that reported an explicit requirement of passage of return-to-sports testing. The athlete must perform a series of dynamic functional activities with resistance from a sport cord such as single-leg squats for 3 min, lateral bounding for 80 s, and forward/backward jogging for 2 min each. He or she is graded on the ability to demonstrate good neuromuscular control of the lower extremity during multi-planar movements that simulated athletic activities [6].

19.5 Patient-Reported Outcome

The rehabilitation protocol efficacy should be assessed using patient-reported outcome. The instruments that are appropriate for use in this patient population and have been recommended to guide therapy progression are Hip Outcome Score (HOS), the International Hip Outcome Tool (iHOT-33/iHOT-12), and the Copenhagen Hip and Groin Outcome Score (HAGOS) [9].

Conclusion

Rehabilitation following hip arthroscopy should be individualized and evaluation based rather than time based. Weight bearing and motion progression are based upon the specific surgical techniques performed. The protocol proposed here is a summary of the current literature together with our experience and the protocols applied in our daily work.

It becomes very important to the clinical evaluation to use scales for initial evaluation and at the end of the rehabilitative path. As we have developed a criteria-driven rehabilitation protocol for safe integration and return to sports after ACL reconstruction, the same must be done following hip arthroscopy.

References

1. Bedi A, Chen N, Robertson W, Kelly BT. The management of labral tears and femoroacetabular impingement of the hip in the young, active patients. Arthoscopy. 2008;24(10):1135–45.
2. Montgomery S, Ngo SS, Hobson T, et al. Trends and demographics in hip arthroscopy in the United States. Arthorscopy. 2013;29(4):661–5.
3. Edelstein J, Ranawat A, Enseki KR, et al. Post-operative guidelines following hip arthroscopy. Curr Rev Muscoloskelet Med. 2012;5(1):15–23.
4. Grzybowski JS, Malloy P, Stegemann C, et al. Rehabilitation following hip arthroscopy—a systematic review. Front Surg. 2015;2:21.
5. Stalzer S, Wahoff M, Scanlan M. Rehabilitation following hip arthroscopy. Clin Sports Med. 2006;25:337–57.
6. Garrison JC, Osler MT, Singleton SB. Rehabilitation after arthroscopy of an acetabular labral tear. N Am J Sports Phys Ther. 2007;2(4):241–50.
7. Neumann G, Mendicuti AD, Zou KH, et al. Prevalence of labral tears and cartilage loss in patients with mechanical symptoms of the hip: evaluation using MR arthrography. Osteoarthr Cartil. 2007;15:909–17.
8. Voight ML, Robinson K, Gill L, Griffin K. Postoperative rehabilitation guidelines for hip arthroscopy in an active population. Sports Health. 2010;2(3):222–30.
9. Enseky KR, Kohlrieser D. Rehabilitation following hip arthroscopy: an evolving process. J Sports Phys Ther. 2014;9(6):813–26.
10. Retchford TH, Crossley KM, Grimaldi A, et al. Can local muscles augment stability in the hip? A narrative literature review. J Muscoloskelet Neuronal Interact. 2013;13(1):1–12.
11. Torry MR, Schenker ML, Martin HD, et al. Neuromuscular hip biomechanics and pathology in the athlete. Clin Sports Med. 2006;25(2):179–97.
12. McCarthy JC, Lee J. Hip arthroscopy: indications, outcomes, and complications. J Bone Joint Surg Am. 2005;87:1138–45.
13. Sahrmann S. Diagnosis and treatment of movement impairment syndromes. St Louis: Mosby; 2002. p. 1–50.

Postsurgical Rehabilitation: Hernioplasty

20

Simona Cerulli

If the definition of sports hernia and its etiopathogenesis and consequently its treatment are already not unambiguous, we have even less indications on rehabilitation program. In fact, if we analyze the bibliography (PubMed, PEDro, Science Direct), items on post hernioplasty rehabilitation are almost absent, compared with publications for the management of inguinal pain after surgery. This is because there is still a need for guidelines and methods of assessments in order to produce scientific papers of value. However, a distinction arises in literature between the conservative approach, recommended for the first 4–6 weeks of the onset of pain, and the surgical approach, recommended for those suffering from chronic groin pain for wall weakness (pain present for at least 6 months). Another distinction is placed according to the interventions the surgeons used for the athlete, i.e., minimal repair technique (see Muschaweck and Berger), mini open mesh approach (LM Brunt), and laparoscopic approach with increasing degrees of complexity and therefore of functional recovery.

Anyway, prior to considering a surgical solution, it's important to give a conservative chance for recovery.

As a lower abdominal muscle strain or a groin injury, the treatment should hold rest, medication as NSAIDs, and physical therapy. About this, it will be important to consider all the other nosological entities that could be present, as pubic osteitis, femoroacetabular impingement, adductor strains, or tendinopathy. In most of the publications about it, training programs including physical therapy, as ultrasound, laser, and electric stimulation [1]. Weir et al. [2] proposed a randomized controlled clinical trial based on previous studies by Holmic et al. and Weir et al. [3]. They compare a new model of therapy (MMT, multimodal treatment) with the current therapy with the highest level of evidence (ET, exercise therapy). Athletes in ET group received instructions on performing 2 weeks of exercises: static adduction against ball soccer between feet or knees lying supine, abdominal contractions in straightforward and oblique directions, combined abdominal sit-ups and hip flexion, balance training on wobble board, and one-foot exercises on sliding board. After 2 weeks of training program, they were instructed on the second training module during at least 4 weeks: leg abduction and adduction exercises, low back extension exercises, one-leg weight pulling adduction/abduction standing, abdominal sit-ups both straightforward and oblique directions, one-leg coordination exercises, and balance training. During these 6 weeks, only cycling was allowed. At the sixth week, athletes can return to running program (slow jogging then straight sprints then cutting). All the programs were for 8 weeks. In the MMT group, the athletes were treated before

S. Cerulli, M.D.
Medicina Fisica e Riabilitazione, Casa di Cura Quisisana/Villa Stuart Sport Clinic, Rome, Italy
e-mail: simona.cerulli@gmail.com

R. Zini et al. (eds.), *Groin Pain Syndrome*, DOI 10.1007/978-3-319-41624-3_20

performing training by a sports medicine physician that works on muscles, checking tensions and stretching adductor muscles (warmed before the manual therapy with paraffin packs for 10 min). The treatment involved even hip circular motion and compression. After this, the athletes have to perform a 5 min warming up every day using slow jogging or cycling. After 15 days of stretching and daily warming up, the athletes if there is no pain are able to start the return to running program as in the ET group. At the end, the results show that the athletes who were treated with the MMT were able to return to sports activities more quickly than the ET group.

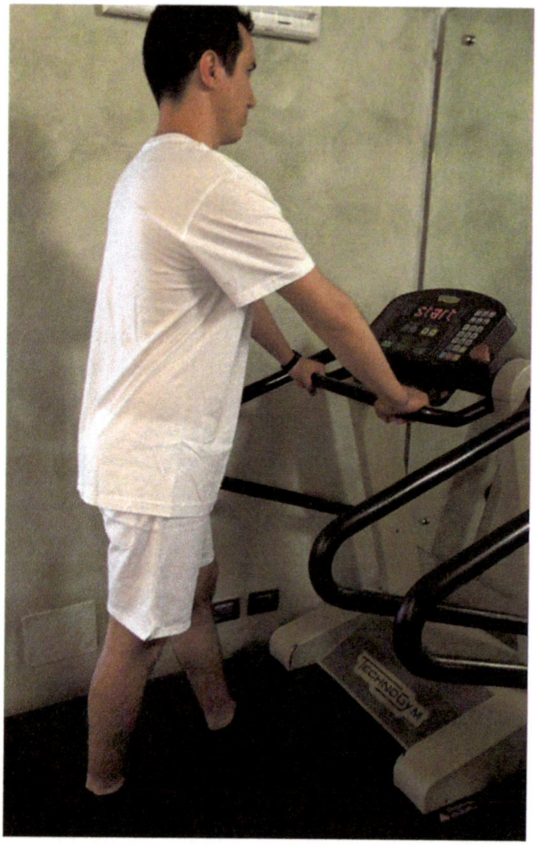

Fig. 20.1 Walking on plane surface

Searching for items on postsurgical programs, the sources are very poor and the phases through which to implement the training were not explained.

Muschaweck suggests [4–6] that all patients have to be discharged on the day of intervention; NSAIDs were used for postsurgical pain relief; patients are allowed to lift up to 20 kg immediately after operation, running and cycling on the second postoperative day (POD), specific training on POD 3–4, and complete their training on POD 5. But in both of items, she doesn't tell us what kind of specific training is allowed to have.

Brunt and Barile in their point of view prefer to use an open tension-free mesh repair: in this way, the floor is reconstructed by suturing the mesh to the transversus aponeurosis and to the inguinal ligament and even resecting the ilioinguinal nerve if entrapped. The postoperative protocol [7] that they suggest is focused on abdominal strengthening, stabilization, and flexibility as well as lower body strength, flexibility, and balance. The program develops into five phases, one per week, starting in the first week with walking for short distance (Fig. 20.1); when the patient is able to walk for 30 min continuously, he can start light hamstrings, quadriceps, calf, and low back stretching as tolerated. At the second week, the patient can start with active hip ROM exercise, walking on incline treadmill (emphasizing full extension on walking gait), backward walking, bike workouts, wall sit with Swiss ball, and abdominal drawing-in maneuver (Figs. 20.2, 20.3, 20.4, and 20.5). At the third

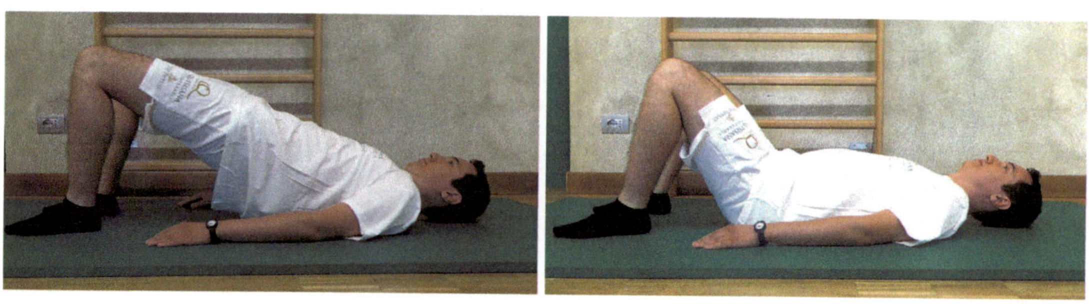

Fig. 20.2 Starting with pelvic tilt (not shown in picture) and bridge on the *right*

Fig. 20.3 Pelvic stabilization with pressure on the ball on the *left* and lifting the leg on the *right*

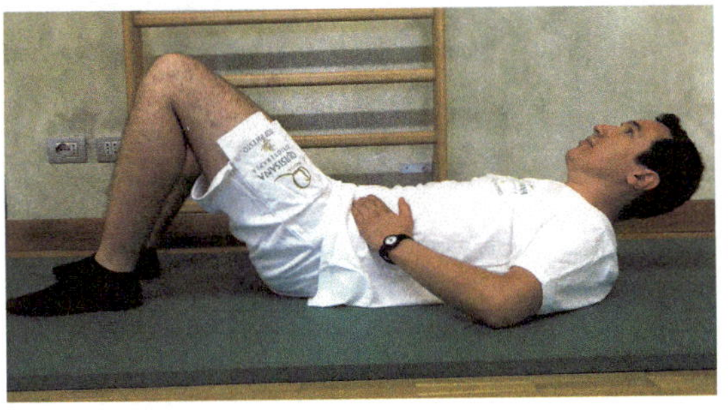

Fig. 20.4 Rectus abdominis isometric contraction

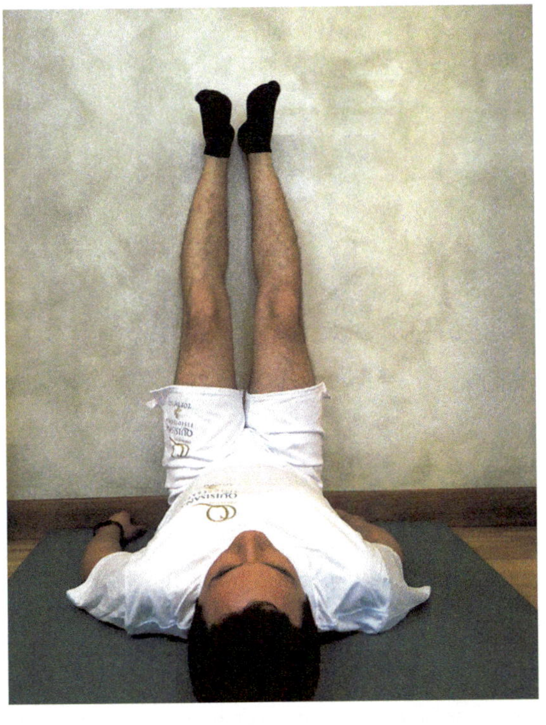

Fig. 20.5 Postural exercises for lumbar spine

week, he has to continue exercise form phase 2 and begin scar mobilization, pool walking, and monitored sports-specific skill activities (ball dribbling, ball hitting progression, skating, and so on). At the fourth week, exercise is increased, with progression in strength, weight, and stabilization so as to arrive at the fifth week during which proper muscle length and abdominal strength are maintained and a core stabilization program is emphasized (Figs. 20.6, 20.7, 20.8, and 20.9). After this period, a medical advice could give the authorization for the reprise of sports activity.

Paajanen et al. [8] give their point of view about the treatment of 60 patients with laparoscopy surgery (endoscopic total extraperitoneal (TEP) mesh placement), demonstrating by their studies that TEP is more effective than nonoperative treatment in athletes. But they didn't tell us anything about postsurgical protocol. It is easier on items [9, 10] to find protocols on postsurgical pain.

Van Veen et al. [11] describe a 6-week post laparoscopic repair protocol. In the first week, walking 5 km/h is allowed; in the second, aquatic training, power walking, stationary cycling, isometric rectus abdominis training, and step-ups are all admitted. During the third through the fifth, the athlete can do training with weight and normal activities within pain-free limits. At the sixth week, the athlete can get unrestricted training.

A more detailed program, and by the way closer to the author's activity in laparoscopic

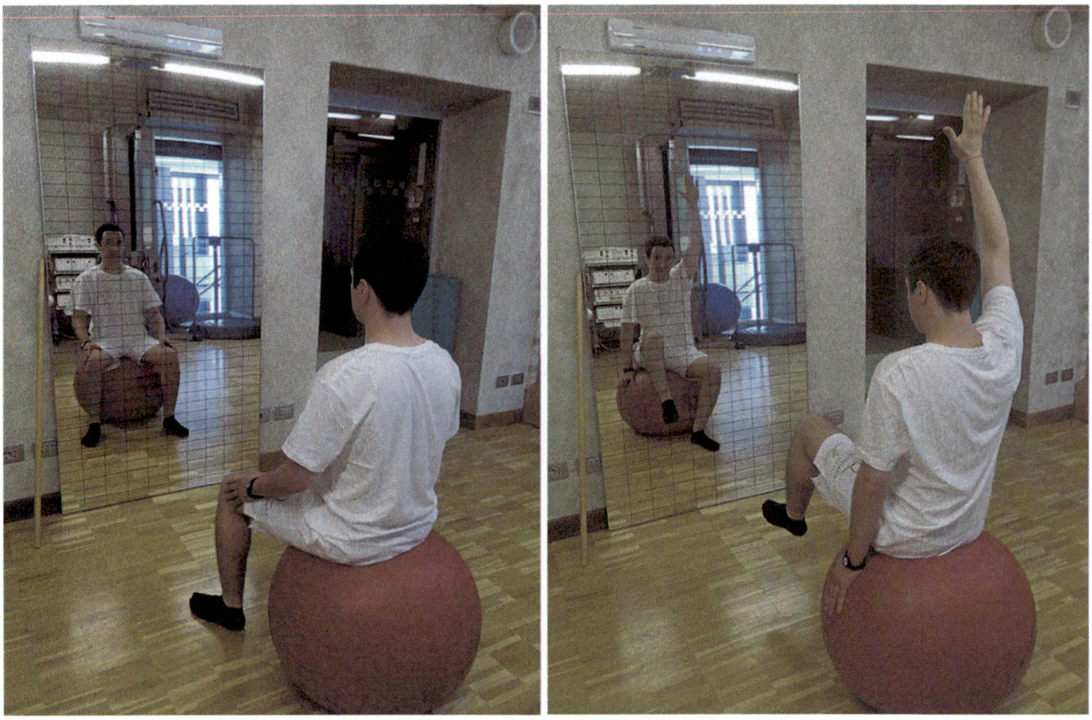

Fig. 20.6 Hip control and core stabilization exercise with mirror feedback

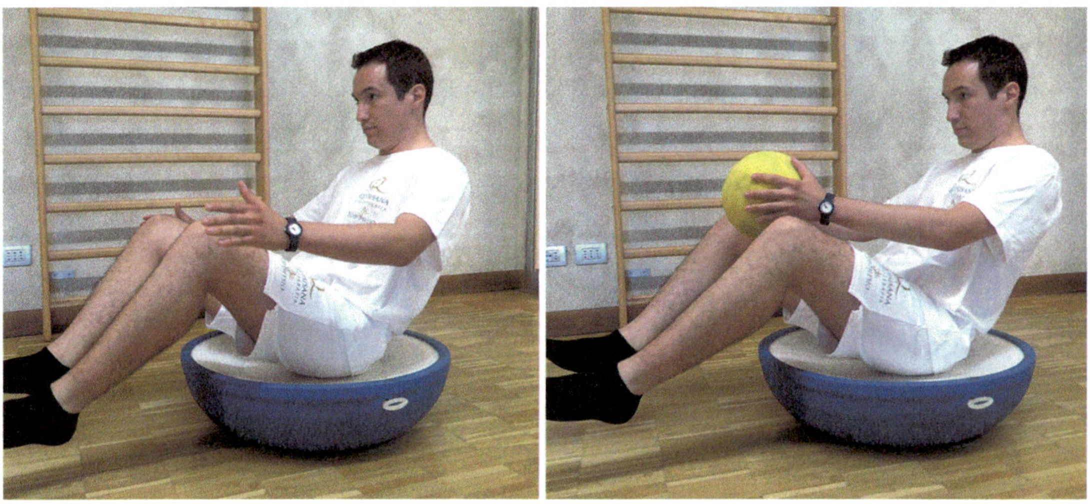

Fig. 20.7 Pelvic control on unstable surface

approach, is proposed by Ellsworth et al. [12]. The whole program takes at least 6–8 weeks. In the first week, wound care, ice, walk in flat surface, and activities of daily living are allowed. During the second and third week, the athlete starts with light resistive exercise in pool, standing closed chain activities for lower extremities and hip, local activation of transversus abdominis, multifidus, iliopsoas, deep hip rotators, deep tissue massage of adductor muscles, and light stretch (lateral trunk, hip extensor, psoas, hamstrings, quadriceps). At the fourth week, core stability program and proprioceptive/balance exercise will start. Scar mobilization over incisional area will

Fig. 20.8 Single leg balance at hip and knee flex; the ball on the *right* makes a better control of the balance

begin. During the successive weeks, the work will increase as far as the cardiovascular activities at the sixth week, when jogging, rope jumping, crossover cariocas, agility and coordination drills, and plyometrics are all allowed.

In a review on 2008, Caudill [13] reports that postsurgical recovery time, suggested as the return to sports activity, for patients who underwent open repair was 17.7 ± 13.1 weeks compared with 6.1 ± 4.5 weeks for laparoscopic repairs.

According to Caudill, several items are needed to have a real postsurgical guideline:

- More standardized patient outcome
- Activity level assessments
- Serial measurement (there's one study [9] in which preoperative and postoperative isometric/isokinetic evaluation is reported) on specific criteria as pain, strength, and functional pain

Fig. 20.9 Trunk and pelvic control in different shoulders placing, on the forearm, on the hand, and on unstable surface

The fact is that the way to approach surgery is quite different between operators, each one with different complications and outcomes, and obviously the rehabilitation has to follow with the limits that we've discussed.

Author's opinion is that the achievement of a univocal way to approach post hernioplasty rehabilitation program is necessary in order to gain an evidence-based protocol.

References

1. Verral GM, Slavotinek JP, Fon GT, Barnes PG. Outcome of conservative management of athletic chronic injury diagnosed as pubic bone stress injury. Am J Sports Med. 2007;35:467–74.
2. Weir A, Jansen JACG, Van de Port IGL, Van de Sande HBA, Tol JL, Backx FJG. Manual or exercise therapy for long-standing adductor related groin pain: a randomised controlled clinical trial. Man Ther. 2011;16:148–54.
3. Weir A, Veger SA, Van de Sande HBA, Bakker EW, de Longe S, Tol JL. A manual therapy technique for chronic adductor-related groin pain in athletes: a case series. Scand J Med Sci Sports. 2009;19:616–20.
4. Muschaweck U, Berger L. Minimal repair technique of sportsman's groin: an innovative open-suture repair to treat chronic inguinal pain. Hernia. 2010a;14:27–33.
5. Muschaweck U, Berger L. Sportsmen's groin—diagnostic approach and treatment with the minimal repair technique: a single center uncontrolled clinical review. Sports Health. 2010b;2:216–21.
6. Minnich JM, Hanks JB, Muschaweck U, Brunt LM, Diduch DR. Sports hernia. Diagnosis and treatment highlighting a minimal repair surgical Technique. Am J Sports Med. 2011;39(6):1341–9.
7. Brunt M. and Barile R. My approach to Athletic Pubalgia. On JWT Byrd. In: Operative hip arthroscopy. New York: Springer; 2013. p. 55–65.
8. Paajanen H, et al. Laparoscopic surgery for chronic groin pain in athletes is more effective than non operative treatment: a randomized clinical trial with magnetic resonance imaging of 60 patients with sportsman's hernia (athletic pubalgia). Surgery. 2011;150(1):99–107.
9. Citfi F, et al. Early-stage quantitative analysis of the effect of laparoscopic versus conventional inguinal hernia repair on physical activity. Chirurgia (bucur). 2015;110(5):451–6.
10. McIntosh A, et al. Evidence-based management of groin hernia in primary care—a systematic review. Fam Pract. 2000;17(5):442–7.
11. Van Veen RN, et al. Successful endoscopic treatment of groin pain athletes. Surg Endosc. 2007;21: 189–93.
12. Ellsworth AA, et al. Athletic pubalgia and associated rehabilitation. Int J Sports Phys Ther. 2014;9(6):774–84.
13. Caudill PH, et al. Sports hernias: a systematic literature review. Br J Sports Med. 2008;42:954–64.

Postsurgical Rehabilitation: Tendon Surgery

21

Rita Guitalti and Maria Teresa Pereira Ruiz

21.1 Introduction

Diagnosis and appropriate management in groin pain are difficult and often attributed to the complex anatomy of the groin area with frequent overlapping extra-articular and intra-articular causes [1]. It is important to know that the 27% of the cases of athletic groin pain can be multifactorial [2, 3].

Until now, the most common cause of groin pain in athletes was considered adductor dysfunction [4], but in a recent review, we can see that the top causes for groin pain can be found in femoroacetabular impingement (FAI) (32%) followed by athletic pubalgia (24%), adductor-related pathology (12%), inguinal pathology (10%), and labral pathology (5%) [5]. The adductors play an important role in acting with the lower abdominal muscles to stabilize the pelvis. The most frequent chronic lesion can be attributed to the tendonitis of the adductor longus muscle and may be caused by an underlying partial tendon rupture resulting from an overuse or strain injury [6]. Various studies have confirmed greater frequency in footballers followed by hockey players [5, 7]. It is important to know that while a lack of flexibility plays an important role in hamstring and quadriceps injury, it is not considered a risk factor for adductor injuries. Instead, previous injury and a history of reduced muscle strength have been identified as risk factors for adductor injuries [8]. Studies have demonstrated a less successful outcome with a rehabilitation program of massage and stretching in chronic adductor injury. Adductor strengthening is essential for the prevention and also for non-operative treatment. It has been demonstrated that a rehabilitation program with strengthening of adductors and pelvis-stabilizing muscles is more effective than a program consisting of local therapy and stretching [9].

Surgical approach in groin pain in athletes remains a significant challenge [10]. When we consider tendon surgery in groin pain, we consider adductor tenotomy.

For a good rehabilitation program, it is essential to know the proximal anatomy of the adductor. Davis et al. in a study in 20 dissected adductors muscles saw the differences in the proximal tendon in females and males. They found statistically significant differences in the vascularization of each of the proximal tendons of the adductor longus (AL) and adductor brevis (AB) with a significant decrease nearer the enthesis, and a distinctive intramuscular tendon was present in both [11]. They found that the AL and AB had a deep tendon extension like the hamstring muscles [12]. Functionally, an intramuscular

R. Guitalti (✉) • M.T. Pereira Ruiz
Isokinetic Sport Rehabilitation Center Milano, Montallegro Clinic, Via Garibaldi 3, Genova, 16124 GE, Italy
e-mail: ritaguitaldi@hotmail.it; info@mtpereirafisiatra.it

R. Zini et al. (eds.), *Groin Pain Syndrome*, DOI 10.1007/978-3-319-41624-3_21

tendon may provide extra strength and stability during muscle contraction and could be especially important in muscle such as the AL whereby the muscle-tendon junction is subjected to large mechanical stresses [13]. When we speak of adductor surgical pathology, there is not much in literature; there is not a consensus neither on the technique nor on the rehabilitation program to follow [5]. In this work therefore, we are trying to do a revision of the surgical techniques and the protocols followed.

21.2 Surgical Techniques and Complications

The consensus of all authors is that the criterion for surgery is a history of long-standing and localized district pain at the origin of the adductor longus muscle, refractory to conservative treatment. Unfortunately, little is published on the surgical treatment when conservative measures fail [14] and the follow-up after surgery of the studies does not exceed 10 years.

In a recent review of surgical diagnoses, investigations and treatment of athletic groin pain the surgical techniques used to treat adductor pathology were 5 in a total of 570 patients: percutaneous adductor longus tenotomy (186 patients), open adductor longus tenotomy (155 patients), unspecified adductor longus tenotomy (62 patients), adductor reattachment with suture anchors (5 patients), and partial adductor release (162 patients) [5]. We think that for the success of a rehabilitation program, it is important to know the characteristics of the patient and the specific surgical technique used. One fundamental point is the collaboration between the surgeon and the physiotherapy team. The postoperative complications included are adductor bruising, scrotal edema, wound infection, monolateral weakness in adduction, and painful scar [14–16].

21.3 Rehabilitation Protocols

Unfortunately, the literature on the rehabilitation protocols after surgery on the adductor longus is little, and there are no studies demonstrating that a protocol is better than another. The majority of the protocols are not very specific and not well described, as well as surgical techniques are not homogeneous, and even some studies do not specify clearly the type of surgical approach made on the adductor [5].

Before illustrating the described protocols, we always recommend to make an assessment at the beginning of treatment and at the end; in fact it would be better to pre-op and at the end of the rehabilitation program. But, if by chance, you get a patient without an assessment, we strongly recommend using the rating scales, either the ones we suggest or any other used by you. We recommend the Groin Disability Score (GDS), which is a new scoring modified from the Oxford Hip Score [17], where patients are asked a series of eight questions with response scores ranging from 0 (best) to 4 (worst) for each, resulting in a total score of 0–32 (Table 21.1). In addition, we advise a ten-point pain visual analog scale (VAS) (0 defining no pain and 10 defining the worst possible pain) and the four-point pain functional classification scales Puffer and Zachazewski [18] (Table 21.2).

Postoperative rehabilitation consists of three separate phases.

21.3.1 Phase I

Phase I occurs during the first week after surgery. Patients are discharged home with full weight bearing using crutches for assistance. The goal is to allow the wound to heal to prevent reattachment of the adductor longus tendon [4].

When a complete tenotomy is performed, some authors advise only regular ice packs 10–20 min several times a day for the first week [14, 15].

Table 21.1 Groin disability score (total 0–32)

During the past 4 weeks	Scoring categories
How would you describe the pain you usually have in your groin?	0. None
	1. Very mild
	2. Mild
	3. Moderate
	4. Severe
Have you been troubled by pain from your groin in bed at night?	0. No nights
	1. Only one or two nights
	2. Some nights
	3. Most nights
	4. Every night
Have you had any sudden, severe pain—"shooting," "stabbing," or "spasms" from your groin?	0. No days
	1. Only 1 or 2 days
	2. Some days
	3. Most days
	4. Every day
Have you been limping when walking because of your groin?	0. Rarely/never
	1. Sometimes or just at first
	2. Often, not just at first
	3. Most of the time
	4. All of the time
For how long have you been able to walk before the pain in your groin becomes severe (with or without a walking aid)?	0. No pain for 30 min or more
	1. 16–30 min
	2. 5–15 min
	3. Around the house only
	4. Unable to walk at all because of the pain
Have you been able to climb a flight of stairs?	0. Yes, easily
	1. With little difficulty
	2. With moderate difficulty
	3. With extreme difficulty
	4. No, impossible
Have you had any trouble getting in and out of a car or using public transportation because of your groin?	0. No trouble at all
	1. Very little trouble
	2. Moderate trouble
	3. Extremely difficult
	4. Impossible to do
How much has pain from your groin interfered with your usual work, including housework?	0. Not at all
	1. A little bi
	2. Moderately
	3. Greatly
	4. Totally

Table 21.2 Functional pain classification scale of Puffer and Zachazewski [18]

Classification	Characteristics
Type 1	Pain after activity only
Type 2	Pain during activity, not restricting performance
Type 3	Pain during activity, restricting performance
Type 4	Chronic, unremitting pain

Fig. 21.1 Stretch the legs (without pain) into wide abduction every 2–3 h

We advise a closed-chain adductor strengthening exercise program starting 2 days after surgery. By postoperative day 3, patients are instructed to stretch the legs (without pain) into wide abduction every 2–3 h (Fig. 21.1). Some advise the use of pillows between the legs when sleeping to maintain abduction [4, 16].

21.3.2 Phase II

This phase occurs during the second to fifth postoperative week depending on the type of surgery performed. The goals for this phase are to prevent reattachment, as well as the formation of excess scar tissue, and to improve flexibility of the adductor muscles. When the closed-chain adductor strengthening exercises could be carried out pain-free, open-chain adductor strengthening exercises were started. These strengthening exercises include standing single-leg abduction, supine single-leg straight-leg lifts, standing single-leg circumduction, and straight-leg sit-ups (Fig. 21.2). All exercises are performed 20 times each, 2–3 times per day if possible. Hydrotherapy was encouraged and was started from the moment the wound was healed. In this phase care is taken to improve the overall core strength and reestablish recruitment patterns of the core musculature abduction [4, 16] (Fig. 21.3). Increasing degrees of adductor resistance were introduced over the following weeks with a similarly gradual reintroduction of walking, straight-line jogging, and agility training.

Fig. 21.2 Increase adductor strengthening exercises

21.3.3 Phase III

This phase begins 6–12 weeks postoperative. The goal of this phase is a return to sports. Sprinting and kicking were attempted before week 7–8. Patients are allowed to return to their respective sports level when they show complete wound healing, full hip range of motion, painless adductor muscle activity, and resolution of their preoperative groin complaints. This typically occurs by 8–10 postoperative weeks [4, 14, 16].

Take-Home Message
There was limited evidence for all the rehabilitation protocols after surgery on the adductor longus tendon. Other treatment options were evaluated in the included studies, as all low-quality studies showed an improvement in time in the intervention groups. Anyway, adductor tenotomy provides good symptomatic and functional improvement in chronic adductor-related groin pain refractory to conservative treatment.

Fig. 21.3 Core stability exercises

References

1. Larson CM, Pierce BR, Giveans MR. Treatment of athletes with symptomatic intra-articular hip pathology and athletic pubalgia/sports hernia: a cases series. Arthroscopy. 2011;27:768–75.
2. Bradshaw CJ, Bundy M, Falvey E. The diagnosis of longstanding groin pain: a prospective clinical cohort study. Br J Sports Med. 2008;42:851–4.
3. Lovell G, Galloway H, Hopkin W, et al. Osteitis pubis and assessment of bone marrow edema at the pubic symphysis with MRI in an elite junior male soccer squad. Clin J Sport Med. 2006;16:117–22.
4. Gill TJ, Carrol KM, Makani A, et al. Surgical technique for treatment of recalcitrant adductor longus tendinopathy. Arthrosc Tech. 2014;3(2):e293–7.
5. De Sa D, Hölmich P, Phillips M, et al. Athletic groin pain: a systematic review of surgical diagnoses, investigations and treatment. Br J Sports Med. 2016;50(19):1181–6.
6. Kalebo P, Karlsson J, Swärd L, Peterson L. Ultrasonography of cronic tendon injuries in the groin. Am J Sports Med. 1992;20:634–9.
7. Ekstrand J, Hagglund M, Walden M. Epidemiology of muscle injuries in professional football (soccer). Am J Sports Med. 2011;38:1226–32.
8. Engebretsen A, Myklebust G, Holme I, et al. Intrinsic risk factor for groin injuries among male soccer players: a prospective cohort study. Am J Sports Med. 2010;38:2051–7.
9. Hölmich P, Uhrskou P, Ulnits L, et al. Effectiveness of active physical training as treatment for long-standing adductor related groin pain in athletes: randomized trial. Lancet. 1999;353:439–43.
10. Hölmich P. Long-standing groin pain in sportspeople fall into three primary patterns, a "clinical entity" approach: a prospective study of 207 patients. Br J Sports Med. 2007;41:247–52.
11. David JA, Stringer MD, Woodley SJ. New insights into the proximal tendons of adductor longus, adductor brevis and gracilis. Br J Sports Med. 2012;46:871–6.
12. Sj W, Mercer SR. Hamstring muscles: architecture and innervation. Cells Tissues Organs. 2005;179:125–41.
13. Tuite DJ, Firegan PJ, Saliaris AP, et al. Anatomy of the proximal musculotendinous junction of adductor longus muscle. Knee Surg Sports Traumatol Arthrosc. 1998;6:134–7.
14. Atkinson HD, Johal P, Falworth MS, et al. Adductor tenotomy: its role in the management of sports-related groin pain. Arch Orthop Trauma Surg. 2010;130(8):965–70.
15. Maffuli N, Loppini M, Longo UG, Denaro V. Bilateral mini-invasive adductor tenotomy for the management of cronic unilateral adductor longus tendinopathy in athletes. Am J Sports Med. 2012;40(8):1880–6.
16. Schilders E, Dimitrakopoulou A, Cooke M, et al. Effectiveness of selective partial adductor release for chronic adductor-related groin pain in professional athletes. Am J Sports Med. 2013;41(3):603–7.
17. Dawson J, Fitzpatrick R, Carr A, Murray D. Questionnaire on the perceptions of patients about total hip replacement. J Bone Joint Surg Br. 1996;78(2):185–90.
18. Puffer JC, Zachazewski JE. Management of overuse injuries. Am Fam Physician. 1988;38:225–32.

Part 8

Conclusion

22 Groin Pain Syndrome: A Final Assessment of the State of the Art

Raul Zini, Piero Volpi, and Gian Nicola Bisciotti

22.1 A Controversial Pathology

The first diagnosis of groin pain is due to Spinelli and dates more than 80 years ago [1]. Since then this disease has never ceased to arouse interpretative and conceptual controversies. From etiopathogenic point of view, we know that the groin pain syndrome (GPS) is more frequent in sport activities like football, hockey, rugby, and distance running [2–11]. However, none of these publications relate the incidence of the injury to the number of licensed athletes into the various activities in question, and most of these studies would be rejected if we follow the minimum criteria of a meta-analysis. In effect, in the literature we find classified different types of GPS according to the type of pathologic lesion and to symptoms that are reported by the patient. For this reason often, an inaccurate diagnosis, leading to inadequate therapeutic interventions, can further lead to a very debilitating medical problem, sometimes forcing the athlete to long suspension of sport activity.

22.2 The Various Classification Attempts

The first attempts of a rational classification are attributable to Durey and Rondineau [2] and Brunet in 1983 [12]. According to the experience of these authors, the GPS in athletes refers to three different anatomo-clinical entities often associated as follows:

1. Parieto-abdominal pathology, affecting the lower part of the anterior abdominal muscles (external and internal oblique muscles and transverse muscle), fascia transversalis, conjoint tendon, and inguinal ligament
2. Adductor muscle pathology mainly affecting the adductor longus and pectineus muscle
3. Pubic symphysis pathology

Then, to this classifications followed by many authors, a certain number of other classifications are more or less interesting and conceptually more or less correct [5, 13, 14].

R. Zini (✉)
Maria Cecilia Hospital GVM Care and Research, Ravenna, Italy
e-mail: raulzini@gmail.com

P. Volpi
Istituto Clinico Humanitas, Rozzano, Milan, Italy

Responsabile Settore Medico F.C. Internazionale, Milan, Italy

G.N. Bisciotti
Qatar Orthopedic and Sport Medicine Hospital, FIFA Center of Excellence, Doha, Qatar

R. Zini et al. (eds.), *Groin Pain Syndrome*, DOI 10.1007/978-3-319-41624-3_22

However, to arrive to a first rational and complete GPS classification, we must to wait 2008 with the classification of Omar et al. [15]. This classification suggests a differential diagnosis of groin pain syndrome based on 37 major diseases, subdivided in ten different categories.

22.3 The Consensus Conference Era

The real conceptual revolution into the GPS classification will come with the era of Consensus Conference, the first of which will be in Manchester in 2013 [16].

The aim of this first consensus conference was to produce a multidisciplinary consensus to determine the current position on the nomenclature, definition, diagnosis, imaging modalities, and management of GPS. The real news was in the fact that for the first time, a group of experts met to create a consensus document. Experts in the diagnosis and management of GPS were invited to participate in a consensus conference held by the British Hernia Society in Manchester on 11–12 October 2012. This team of experts included a physiotherapist, a musculoskeletal radiologist, and surgeons with a proven track record of expertise in this field. One of the most important results of the consensus conference was the adoption of the term "inguinal disruption" (ID) agreed as the preferred nomenclature. It was established that ID is a common condition where no true hernia exists. It should be managed through a multidisciplinary approach to ensure consistent standards and outcomes are achieved. It was the first time that the concept of "multidisciplinary approach" was strongly stressed.

After the Manchester Consensus Conference followed soon the Consensus Conference of Doha [17]. The aim of this second consensus conference was to agree on a standard terminology, along with accompanying definitions. At the end the expert participating to the consensus established that the classification system has three major subheadings of groin pain in athletes, i.e.:

1. Defined clinical entities for groin pain: adductor-related, iliopsoas-related, inguinal-related, and pubic-related groin pain
2. Hip-related groin pain
3. Other causes of groin pain in athletes

The third consensus in the order of time was the "Groin Pain Syndrome Italian Consensus Conference on terminology, clinical evaluation and imaging assessment in groin pain in athlete," held in Milan in February 2016 [18], concerning which we talk extensively in Chap. 1. One of the most important results of this conference was the formulation of a flow chart as previously hoped, as "future direction," into the Manchester Conference. This is the first time that a flow chart for the decision-making process in GPS is presented. The flow chart is showed in Fig. 22.1.

22.4 Future Directions

Future developments of the GPS study are undoubtedly entrusted to the development of the consensus conference and then of the multidisciplinary study. In fact, only by involving all the professional figures concerned in GPS management, we can hope to progress in its knowledge. Specifically, many efforts should be made investigating the relationship between FAI and inguinal pathologies. Furthermore, the study of inguinal pathologies in which a true inguinal hernia is not present (in which, therefore, it may be suspected an injury at conjoined tendon, inguinal ligament, aponeurosis of the external oblique muscle, a nerve entrapment, etc.) represents a very interesting field of study.

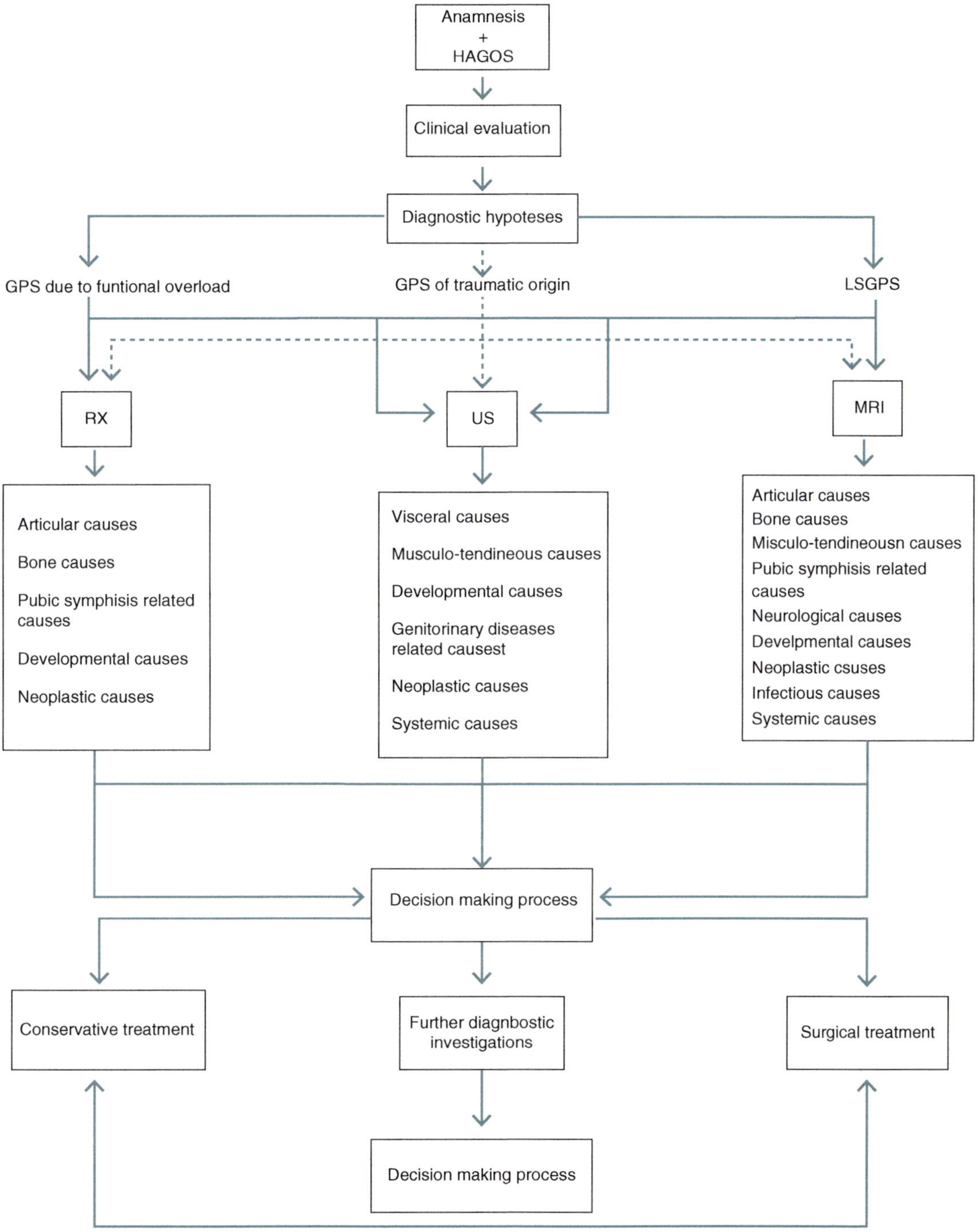

Fig. 22.1 Flow chart based on the results of the "Groin Pain Syndrome Italian Consensus Conference on terminology, clinical evaluation and imaging assessment in groin pain in athlete." After the anamnesis and the clinical evaluation, the patient performs the imaging evaluation. The decision-making process is based on the results of clinical and imaging evaluations. In case of GPS of traumatic origin (as explained in guidelines), the possibility of choice among the various imaging tests is indicated in the flow chart with the *dashed line*. In the case in which it is possible to have a diagnosis, the patient may be advised for a conservative or surgical treatment. In the case in which a diagnosis is not reached, the patient may be advised for further diagnostic investigations (i.e., blood tests, urine test, CT, scintigraphy, etc.) in order to obtain diagnosis and decide the treatment path

References

1. Spinelli A. Una nuova malattia sportiva: la pubalgia degli schermitori. Ortho Trauma App Mot. 1932;4:111.
2. Durey A, Rodineau J. Les lésions pubiennes des sportif. Ann Med Phys. 1976;9:282–91.
3. Durey A. Modifications radiologiques microtraumatiques du pubis. Micro-traumatologie du sport. Masson. 1987;15:185–92.
4. Arezky N, Zerguini Y, Mekhaldi A, Zerdani S, Massen R, Bouras R. La maladie pubienne chez le sportif. Priorité au traitement médical. J Traumatol Sport. 1991;8:91–7.
5. Le Gall F. La pubalgie du sportif. A propos de 214 cas. Thèse en médicine Université de Rennes; 1993.
6. Gilmore J. Groin pain in the soccer athlete: fact, fiction and treatment. Clin Sports Med. 1998;17:787–93.
7. Gibbon WW. Groin pain in professional soccer players: a comparison of England and the rest of Western Europe. Br J Sports Med. 1999;33:435.
8. Ekstrand J, Ringborg S. Surgery versus conservative treatment in soccer players with chronic groin pain: a prospective randomised study in soccer players. Eur J Sports Traumatol Relat Res. 2001;23(4):141–5.
9. Karlsson MK, Dahan R, Magnusson H, Nyquist F, Rosengren BE. Groin pain and soccer players: male versus female occurrence. J Sports Med Phys Fitness. 2014;54(4):487–93.
10. Rankin AT, Bleakley CM, Cullen M. Hip joint pathology as a leading cause of groin pain in the sporting population: a 6-year review of 894 cases. Am J Sports Med. 2015;43(7):1698–703.
11. De Sa D, Hölmich P, Phillips M, Heaven S, Simunovic M, Philippon MJ, Ayeni OR. Athletic groin pain: a systematic reviewof surgical diagnoses, investigation and treatment. Br J Sports Med. 2016;50(19):1181–6.
12. Brunet B. La pubalgie, un syndrome "fourre tout". Thèse Med. Université de Lyon 20; 1983.
13. Bouvard M, Dorochenko P, Lanusse P, Duraffour H. La pubalgie du sportif- stratégie thérapeutique. J Traumatol Sport. 2004;21:146–63.
14. Puig PL, Trouve P, Savalli L. Pubalgia: from diagnosis to return to the sport field. Ann Readapt Med Phys. 2004;47(6):356–64.
15. Omar IM, Zoga AC, Kavanagh EC, Koulouris G, Bergin D, Gopez AG, Morrison WB, Meyers WC. Athletic pubalgia and "sports hernia": optimal MR imaging technique and findings. Radiographics. 2008;28(5):1415–38.
16. Sheen AJ, Stephenson BM, Lloyd DM, Robinson P, Fevre D, Paajanen H, de Beaux A, Kingsnorth A, Gilmore OJ, Bennett D, Maclennan I, O'Dwyer P, Sanders D, Kurzer M. 'Treatment of the sportsman's groin': British Hernia Society's 2014 position statement based on the Manchester Consensus Conference. Br J Sports Med. 2014;48(14):1079–87.
17. Weir A, Brukner P, Delahunt E, Ekstrand J, Griffin D, Khan KM, Lovell G, Meyers WC, Muschaweck U, Orchard J, Paajanen H, Philippon M, Reboul G, Robinson P, Schache AG, Schilders E, Serner A, Silvers H, Thorborg K, Tyler T, Verrall G, de Vos RJ, Vuckovic Z, Hölmich P. Doha agreement meeting on terminology and definitions in groin pain in athletes. Br J Sports Med. 2015;49(12):768–74.
18. Bisciotti GN, Volpi P, Zini R, Auci A, Di Marzo F et al. Groin Pain Syndrome Italian Consensus Conference on terminology clinical evaluation and imaging in groin pain in athlete. BMJ Open Sport Exerc Med. 2016;2:e00142. doi: 1136/bmjsem-2016-00142